Praise for

Why Science Is Wrong...About Almost Everything

"With heart, candidness and humanity, Alex yanks from the realm of academic abstraction the most important questions of our times: What is science-as-you-know-it telling us about what we are and what reality is? Do its answers stand to reason and empirical fact? In making his surprising—sometimes even alarming—case that they don't, Alex shows that all is not as it seems. His argument is imbued with such urgency, relevance, and aliveness that anyone who senses the magnitude of the mystery we live in should read this book immediately."
— Bernardo Kastrup, Ph.D., author of *Why Materialism Is Baloney*

"There's nothing wrong with science itself. But there's a great deal wrong with how some scientists stubbornly perpetuate myths known to be wrong. Tsakiris learned this by interviewing scientists on both sides of controversial topics and seeing first hand how self-proclaimed 'skeptics' often wildly distort and denigrate discoveries just because they challenge prevailing theories."
— Dean Radin, Chief Scientist at the Institute of Noetic Sciences and author of *Supernormal*

"What do we really know about consciousness? This is one of humankind's most important and debated questions. *Why Science is Wrong... About Almost Everything* brings a fresh and exciting perspective to this ancient question. Alex Tsakiris presents an impressive compilation of his discussions with some of the world's leading scholars on the issue of consciousness and science. Alex masterfully demonstrates that consciousness cannot be fully explained by known physical brain function. Multiple interviews confirm that what many leading materialist scientists accept as factual is indeed wrong. This book is well written and enthusiastically recommended."
— Jeffrey Long, MD, author of *Evidence of the Afterlife*

"In his bawdy, baddass style, *Skeptiko* pioneer Alex Tsakiris hunts down the world's leading consciousness researchers to challenge and skewer outdated modes of scientific thinking. Along the way, he doesn't merely show us what's wrong with science, but what's right about all of us."
— Jonathan Talat Phillips, Professional Psychonaut and author of *The Electric Jesus*

WHY SCIENCE IS WRONG
...ABOUT ALMOST EVERYTHING

By Alex Tsakiris

ANOMALIST BOOKS
*San Antonio * Charlottesville*

An Original Publication of ANOMALIST BOOKS

Why Science Is Wrong...About Almost Everything
Copyright © 2014 by Alex Tsakiris
ISBN: 978-1938398315

All rights reserved, including the right to reproduce this book or portions thereof in any form whatsoever.

The opinions expressed in this book are those of the author and do not necessarily reflect the views of Anomalist Books, LLC.

Book design by Seale Studios

Cover design by Danijela Mijailović
Cover image © Brett Lamb/Bigstock.com

For information, go to AnomalistBooks.com, or write to:
Anomalist Books, 5150 Broadway #108, San Antonio, TX 78209

CONTENTS

FOREWORD by Dr. Rupert Sheldrake xi

INTRODUCTION 1
SCIENCE AT THE TIPPING POINT
   Skeptiko.com 3
   It Helps Talking to the Experts 4

CHAPTER 1 5
WRONG ABOUT CONSCIOUSNESS...SAY "HELLO"
   Dr. B. Allan Wallace 7
   Dr. Michael Graziano 8
   Dr. Patricia Churchland 12

CHAPTER 2 19
WRONG ABOUT QUANTUM PHYSICS...IF A TREE FALLS...
   Dr. Mario Beauregard 20

CHAPTER 3 25
WRONG ABOUT SCIENCE...AND ITS DOPEY CREED
   Matt Dillahunty 27

CHAPTER 4 33
WRONG ABOUT NEAR DEATH EXPERIENCE...
   Scientific Consensus Not Enough For True Non-Believers 34
   Shoddy Research Exposed 36
   An Unfair Fight 37

CHAPTER 5 53
WRONG ABOUT PSYCHICS AND MEDIUMS (Part1)
   Dr. Julie Beischel 54

CHAPTER 6 67
WRONG ABOUT PSYCHICS AND MEDIUMS (Part 2)
   Marilynn Hughes 68
   Megan 70
   Michelle 76

CHAPTER 7 79
WRONG ABOUT TELEPATHY...DOGS THAT KNOW
   My Replication of Dogs That Know 80
   Rupert Sheldrake 82

CHAPTER 8 89
WRONG ABOUT PSYCHIC DETECTIVES
   Ben Radford 90
   We Speak to the Homicide Detectives 99
   Just Ignore the Facts 103

CHAPTER 9 105
WRONG ABOUT HEALING & MEDICINE
   Dr. William Bengston 106
   Bernadette Doran 114

CHAPTER 10 123
WRONG ABOUT ATHEISM
   Dr. Gary Marcus 124
   Dialoging With Atheists 126

CHAPTER 11 133
WRONG ABOUT EVOLUTION...AND CHARLES DARWIN
   Michael Flannery 134
   Flannery on Dr. Jerry Coyne 140
   Roy Davies 143

CHAPTER 12 155
THE END OF SCIENCE-AS-WE-KNOW-IT
  Bernardo Kastrup 156
  Dr. Jeffrey Schwartz 159

CHAPTER 13 163
SCIENCE OVER THE TIPPING POINT
  Tim Freke 164
  Consciousness Is Not Meaningless 168

APPENDIX 169
  Dr. Dean Radin's Meta-Analysis of Presentiment
  by John Maguire

NOTES 173

INDEX 177

Author's Note

Some audio interview transcripts have edited for clarity and size limitations. Complete audio versions of the interviews are available at Skeptiko.com.

FOREWORD

Alex Tsakiris is a fearless investigator. He is also persistent and independent-minded. I have known him since he first started his *Skeptiko* podcast and he has interviewed me several times for it. What struck me right from the outset was that he wanted to hear both sides of every argument. For example he wanted to hear what I had to say, and he wanted to hear what my critics had to say, and he wanted to find out how valid their criticisms were. He dug deeper until he could find out the reasons for our differences. In this book, which contains highlights from his investigations on *Skeptiko*, he explores a range of controversial phenomena, looking both at what investigators of these phenomena claim and also at what the skeptics say.

When Alex started his enquiries, he expected that the leaders of organized skepticism would have strong and persuasive arguments, but he soon found they did not. The weakness of their case shows up very strongly in this book. Of course, they provide a useful service in exposing frauds and charlatans, but a strong ideological commitment forces them to deny all evidence that does not fit into their worldview. If outright denial does not work, then they have to muddy the water and create confusion rather than clarity.

I found this book compelling reading, a page-turner. I particularly enjoy the way Alex followed his enquiries wherever they led, including working with the skeptic Ben Radford on an enquiry into information from psychics that helped to solve crimes. When Ben questioned some of the evidence, Alex called the detectives who had been handling the cases, so that he and Ben could together clear the point up by speaking to them directly.

In spite of my appreciation of Alex's investigative skills, and his bravery and commitment to truth, I think the title of this book goes too far. Science is not wrong about almost everything; it is right about a great many things, or right enough. Everyone is appropriately impressed by computers, smartphones, the internet, jet planes, hip replacement surgery, antibiotics, solar panels, and many other technological inventions that enrich and sustain our lives. Science is right about the existence of galaxies beyond our own, about the structures of molecules, about the existence of fossils, about low-temperature superconductivity, and many, many other things. What it's seriously wrong about is the nature of life and consciousness. The materialist theory cannot account for the existence of minds, and also, in my view, gives a very inadequate understanding the nature of life. It would have been better to call this book *Why Science Is Wrong About Consciousness*, or *Why Science Is Wrong About Life and Consciousness*. I told Alex so, but he wanted to stick to his original title.

Alex and his interviewees deal with deep questions at the very frontiers of scientific understanding, in areas where the sciences are being inhibited by fear, dogmatism, and disinformation. I hope this book will serve to clear the ground and lead to a more productive discussion of important questions that rightly interest many people.

Rupert Sheldrake
London, October 2014

INTRODUCTION: SCIENCE AT THE TIPPING POINT

Neither science nor rationality are universal measures of excellence... Science is neither a single tradition, nor the best tradition there is, except for people who have become accustomed to its presence, its advantages, and disadvantages. In a democracy it should be separated from the state just as churches are now separated from the state.[1] — Philosopher Paul Feyerabend

The pause is brief, only a second or so, but I'm panicking. I'm sitting in the patio next to my kitchen—a wonderfully calming space I've converted into a studio for the science-themed podcast I'm recording. But I'm not calm right now. I've asked my question and my mind is racing as I wait for an answer. This could get ugly.

Dr. Dean Radin is on the other end of the recorded Skype call. He's one of the world's leading parapsychology researchers, and rather than offer up polite banter about his bestselling books, I've pressed him with a tough question about his competence as a researcher. I'm not a professional scientist. I have no training as a broadcaster. And to be honest, I haven't gotten all the way through Radin's book, *Entangled Minds*. But none of that matters now. I've asked Radin to respond to some specific claims by one of his harshest critics. I have to see where this goes. Radin begins in a soft-spoken, measured tone:

"It's interesting; he claims I do a lot of studies and don't repeat them, and the very next thing he says is that I repeated the presentiment experiment a number of times."[2]

Radin is responding to claims made by University of Oregon psychologist Dr. Ray Hyman, a respected scientist with a stellar academic background and long list of peer-reviewed publications. Hyman attacked Radin during an interview with Yale neurologist and self-proclaimed "skeptic" Dr. Steven Novella. They were discussing Radin's peer-reviewed research into the nature of time and consciousness. Radin's research posed a fundamental question: *When do we know what we know*? What Radin suspected, and what the data ended up revealing, is we sometimes know things are going to happen *before* they occur.

Radin discovered this by asking test subjects to stare at a blank screen and wait for an image to be displayed. During this time he measured their physiological response to the image. Sometimes he measured galvanic skin response, other times he measured pupil dilation or brain activity. But the goal was always to see if there was a detectable physiological reaction *before* the image appeared. Surprisingly, he did find such a reaction, particularly when troubling or extremely stimulating images were displayed.[3]

Radin had published his results in a peer-reviewed scientific journal. He had also replicated his work by repeating the experiment a number of times to make sure his results were consistent. He even collaborated with other independent researchers in labs throughout the world who were interested in replicating his results (as of this writing Radin's presentiment experiments have been successfully replicated over 25 times in 7 different laboratories). But during the interview with fellow skeptic Steve Novella, Hyman seemed unwilling to even consider such research. And, he saved his sharpest punches for Radin himself. Hyman snickered at Radin's competence and charged Radin with making basic mistakes in collecting and interpreting his data. He even went so far as to say that Radin was "changing his correc-

Introduction: Science at the Tipping Point

tions" in order to "get what [he] wanted from the data."
Before I explain to you exactly how this story ends (spoiler alert: Hyman is shamefully wrong, and Radin's research remains unchallenged; see Appendix), let me back up and explain why a non-scientist entrepreneur such as myself was using an internet radio-show to jump into such an obscure parapsychology debate.

Skeptiko.com

I started the *Skeptiko* podcast in 2007 with the lofty goal of getting at "the truth." The "who are we and why are we here?" truth. I wasn't getting the answers from the books I was reading or interviews I was listening to, so I decided to ask the experts myself. The *Skeptiko* podcast would be my vehicle, my excuse for dialoguing with leading researchers and experts.

I didn't expect *Skeptiko* to reach a large audience. I just wanted the opportunity to talk to the experts. My motto became: *Follow the data, wherever it leads.*

Fortunately, I had stumbled across the work of a maverick biologist named Dr. Rupert Sheldrake. As a highly regarded Cambridge-trained scientist, Sheldrake had tackled these sorts of big-picture questions over the years through some ingeniously designed experiments.

I was even more fortunate when Sheldrake responded to my email queries and encouraged me to investigate further. While I knew where I wanted to go, I didn't know exactly how to get there. I also knew that even if I approached these questions from a purely scientific standpoint, I would inevitably be drawn into broader culture war debates surrounding science and religion. I was looking forward to that as well.

Although I had been raised within the Christian tradition and had a strong cultural attachment to the Greek Orthodox Church (especially the Greek food I was brought up on), I started to question the validity of Bible stories soon after entering college. By the time I was an adult, I wasn't inclined to give much weight to religious explanations of scientific mysteries. At the same time, after spending a couple of years as a research associate at the University of Arizona in pursuit of a Ph.D. in artificial intelligence, and then starting and running my own IT firm, I was becoming less intimidated by academics who offer more credentials than intellectual muscle.

So, as I sat on the other end of the Skype call, listening to Dean Radin systematically tear apart Ray Hyman's claims of "data mining" and inadequate baseline control, I got my first taste of what I was searching for. While I couldn't be sure Radin was giving me the right answers, he was definitely showing me why I might want to be a little more discerning when it came to the opinions of some scientific experts.

It Helps Talking to Experts

Fast-forward seven years. I've interviewed over 200 guests, including many bestselling authors and dozens of world-class academics and thinkers. The *Skeptiko* show I started has inspired millions of downloads and scores of gratifying emails from folks around the world who have shared this journey with me.

Throughout this book I'll share with you what I've learned through my interviews with some of the world's leading consciousness researchers and thinkers. In doing so, I'll reveal what the best research is saying about "big picture" science questions and how they affect our understanding of the limits of science in general.

And because I'd hate to make you wait, I'll give you the condensed version here. Science-as-we-know-it is an emperor-with-no-clothes-on proposition. It mesmerizes us with flashy trinkets, while failing at its core mission of leading us toward self-discovery. Science is wrong about almost everything because science depends on *YOU* not existing. Science is wrong because it assumes your consciousness (i.e. your mind…your awareness…the voice inside your head) is an illusion—and it's not!

CHAPTER 1: WRONG ABOUT CONSCIOUSNESS... SAY "HELLO"

If physical science, whatever it may have to say about the origin of life, leaves us necessarily in the dark about consciousness, that shows that it cannot provide the basic form of intelligibility for this world. There must be a very different way in which things as they are make sense, and that includes the physical world, since the problem cannot be quarantined in the mind.[4] — Philosopher Thomas Nagel

Thank goodness for public television. PBS Cincinnati gave me my first lesson in consciousness. I was just out of college and enjoying a day off from work when I tuned into a strange show called *Lilias Yoga*. At the end of the program, Lilias led me and the rest of her viewers through a guided meditation: "Breathe softly and quiet that voice inside your head," Lilias instructed. Until that moment, I never realized there *was* a "little voice inside my head." I had never really thought about my consciousness in that way. As it turns out, questions about "the little voice inside our head" are not only central to yoga classes, but also to science.

What is consciousness? Let's simplify. Here's an exercise to try. Don't worry, it's an easy one. Take a deep breath and quiet your mind a bit. Now say, "Hello." I know it seems a little strange, but use that voice inside your head to say "Hello" to yourself. Did you hear it? Did you hear YOU say "hello" to YOU? Of course you did. The YOU that heard "hello" is the essence of consciousness.

According to our present, science-as-we-know-it worldview, there is only one acceptable explanation for the origins of this consciousness. Science insists your consciousness is solely and completely a product of your brain. To allow for anything outside your physical brain to be involved in human consciousness would fundamentally alter science. So, before I show you how time and again we have encountered solid evidence suggesting consciousness is something more than just the electrochemical firing of your brain, let me show you why science-as-we-know-it desperately needs this not to be so.

Here's a simple science question: what is the boiling point of water? Short answer: 212° F, or 100°C. But wait, it's not quite that simple. As you probably know, the boiling point of water depends on atmospheric pressure, which changes according to elevation. Water boils at 212° F at sea level, but it boils at a lower temperature in the mountains. The boiling point also depends on the purity of the water. Salty water boils at a higher temperature than pure water. Unless we account for these factors we can't properly answer our science question. In other words, *unless we know all the variables involved, we can't measure the result.*

With that in mind, let's return to the question of consciousness. If my consciousness is solely a product of my brain, I don't have anything to worry about in regards to my water boiling experiment. The world I'm trying to measure is out there, and I'm in here. Everything is neatly organized within this framework. But if my consciousness extends beyond my physical brain, then I have to ask whether my consciousness (i.e. that little voice inside my head) has influenced this world I'm trying to measure. For example, if I concentrate/meditate/pray really hard, can I make this water boil at a lower temperature?

Okay, forget about the boiling water experiment for a minute. Can I use my consciousness to make a plant grow faster?[5] Can I use prayer (i.e. directed consciousness) to make a hospital patient heal sooner?[6] Can I silently extend my consciousness to another person and give them the answer to a question they're seeking? There is no end to these kinds of questions, but I'll add one more. If the voice inside my head can be shown to do any of this, can a scientist ever be sure about what s/he is measuring? The answer is no.

If my consciousness is something—*anything*—other than a product of my brain, then science is out of business until it figures out exactly how my consciousness interacts with this world. If my consciousness is more than my physical brain, then consciousness is the X-factor in every science experiment. It's the asterisk in the footnotes that says, "We came as close as we could, but we had to leave out consciousness in order to make our numbers work."

Dr. B. Allan Wallace

You might think the role of consciousness in science is well understood by most scientists, but it's not. In fact, the biggest obstacle you're likely to encounter while reading this book is the "How can this be?" question. That is, how can all these otherwise smart people be wrong about such obvious issues?

One of the first people I encountered who was both smart enough and brave enough to tackle this question was Dr. B. Allan Wallace. Although he received his Ph.D. in religious studies at Stanford and is best known as a Buddhist scholar, Wallace is also one of the clearest thinking researchers on consciousness with many published, peer-reviewed papers on the topic. It was during an interview with him in 2009 (*Skeptiko* #63) when I was introduced to science's current position on consciousness:

Dr. B. Allan Wallace: *It's absurd! Scientists have yet to explain* the nature of consciousness. They have no means of objectively detecting it. They have not identified its necessary and sufficient causes. And yet, they ask us to wager everything on their belief that consciousness is solely a product of the brain.

One of the things I took away from my interview with Dr. Wallace was his point about the still unidentified "necessary and sufficient causes" of consciousness. This is a big deal in science. If we return to the water boiling experiment, if a scientist were to say s/he didn't know the necessary and sufficient causes of boiling water, we'd conclude they didn't know much about this water-boiling phenomenon. We wouldn't be interested in complicated theories about how water boiled on gas versus electric stoves. We'd insist they go back and work out the basics—what does it take to make water boil?

The same goes for consciousness. Before we accept science's proclamation that consciousness is solely a product of my brain, we must make sure they've covered the basics. They have to be able to explain what is necessary and sufficient to cause consciousness.

Dr. Michael Graziano

I've posed these basic questions about consciousness to some respected scientists at prestigious universities and have been surprised at what I found. Here's an excerpt from my 2014 interview (*Skeptiko* #246) with Princeton University neuroscientist and author Dr. Michael Graziano. Dr. Graziano came to *Skeptiko* promoting a new theory of human consciousness presented in his book, *Consciousness and the Social Brain*. Even though Graziano is a smart guy, teaching at one of our nation's finest universities, he didn't seem to have a very good handle on the necessary and sufficient causes of consciousness. I began by asking him about his theory.

Dr. Graziano: *...the book really lays out theory and it's an entirely rationalist theory. It's a scientific theory of consciousness. And in this way it diverges from a lot of other theories of consciousness that are out there...*

Wrong about Consciousness... Say "Hello"

Me: *...Michael, aren't you just jumping past the hard stuff? I mean, here's the deal—there is this thing called consciousness, so you can say you're assuming it's magic or you're assuming it's not, but we still have to deal with it. So aren't we just jumping past the existence of consciousness? And if you want to go scientific... let's start with the first step. Okay, Dr. Graziano, tell us what's necessary and sufficient to create consciousness. That would be like a first logic, rationalist kind of thing. What's necessary and sufficient to create human consciousness?*

Dr. Graziano: Well, one way to put it, and I have often used this example as it kind of nicely encapsulates our approach, and it is certainly totally different from the perspective that you outlined that I think a lot of people take, so here is an example—I had a friend who was a psychologist and he told me about a patient of his. And this patient had a delusion; he thought he had a squirrel in his head. And that's a little odd, but people have odd delusions and it's not that unusual. Anyway, he was certain of it, and you could not convince him of it otherwise. He was fixed on this delusion and he knew it to be true. Now, you could tell him that's illogical, and he would say, yeah, that's okay, but there are things in the universe that transcend logic. You could not argue him out of it.

So there were two directions you could take in trying to explain this phenomenon. One would be to ask: how does his brain produce a squirrel? How did the neurons secrete the squirrel? Now, that would be a very unproductive approach. And another approach would be to say: how does his brain construct that self-description? And how does it arrive at such certainty that the description is correct? And how does the brain not know that it's a self-description? Now, those things you can get at from an objective point of view. You can answer those questions.

And in effect, I think you could replace the word "squirrel" with the word "awareness" and I think that the whole thing is exactly encapsulated. I think almost all approaches to consciousness take the first direction, how does the brain produce a squirrel—it doesn't.

Me: I guess I would roll back to the first assumption, which is embedded in that, which is that a well-functioning brain creates consciousness. Or even before that, evolution creates brain, brain creates consciousness. And I would say let's start there because you are assuming that. You are saying that is one thing that is necessary to create consciousness, a well-functioning brain.

Dr. Graziano: I do not think brain creates consciousness. I think brains create information and the information informs us of something that is actually not physically correct. And as a result we arrive at this conclusion that we have consciousness and we're captive to that. Brains only know the information available to them, so we're captive to the information generated by our brains, generated within their models. We're captive to that.

So if a brain has a model that tells it, "'you're an elephant," or "You have a squirrel in your head," that brain is captive to that information. It can't escape that. And likewise if brains have information in them that says, "You have this inner experience," then we're captive to that information. I don't think we need to explain how brains produce the inner experience. The thing to be explained is how, and for what adaptive advantage, brains construct the information that they have an inner experience.

Me: But don't we kind of get into a matter of semantics there? I mean, you are assuming that a brain, a well-functioning brain, is necessary to produce everything else in that chain. You say the models produce the information that produces the consciousness, but you're starting with a well-functioning brain. So that is a necessary part of the equation for you, right?

Dr. Graziano: Yes, but I would not buy the word "produce" attached to the word "consciousness." I don't think brains produce consciousness any more than that delusional man's brain produced a squirrel.

Me: Again, that's kind of semantics, right? You are saying brains

produce models, brains process information, and that creates [consciousness]—again, that is a game of semantics in terms of at the end we have this experience we all call consciousness...
One of the reasons conversations like this get stuck in a loop is "group think." Dr. Graziano and his colleagues are so committed to the idea that consciousness is an illusion of the brain that it's become an unexamined assumption in their theory. This leads to the kind of strange ideas you just read where one person's delusion is evidence that all conscious experience is a delusion.

One interesting part of this interview came when I asked Dr. Graziano about evidence suggesting consciousness may persist in the absence of a well-functioning brain, like in the case of near-death experiences (NDEs). During my preparation for the interview I noticed Graziano had never addressed near-death experience research in any of his writings. This seemed strange. NDE research is a hot topic among consciousness researchers, yet he seemed to have no opinion on the subject.

I asked him about a well-known medical paper on near-death experience authored by Dr. Pim van Lommel and published in one of the world's leading medical journals, *The Lancet*.[7] As you'll read, Graziano didn't seem to understand the obvious difference between near-death experience research and research of psychic phenomena:

Dr. Michael Graziano: *I don't know any serious or reputable scientists who have bothered to do experiments to refute psychic phenomena...*

Me: *We're talking about [NDE research published in] The Lancet.*

Dr. Graziano: *Seeing into the future.*

Me: *We're talking about* The Lancet*. We're talking about someone who is a cardiologist for 20 years and you're countering with psychic phenomena?*

Dr. Graziano: *Yes, exactly. That's what it is. It's a psychic phenomenon.*

Me: But you've never examined any of the literature.

Dr. Graziano: *I have read a certain amount of it.*

Me: What have you read? Tell me one that you've read.

Dr. Graziano: *There is a large [body of] literature on astrology that I wouldn't bother to read as an astronomer.*

Me: You can't cite one piece of near-death experience science literature you've read. I mentioned Bruce Greyson at the University of Virginia. I cited three or four people who are well-respected academics. Peter Fenwick, Sam Parnia at Cornell. And you want to talk about psychics. So tell me one that you've read, Michael, one that you've read?

Dr. Graziano: *See, what you're doing is appealing to authority, I'm sorry. It doesn't interest me.*

Me: Appealing to authority!? I'm talking to a Princeton neuroscientist—you are the authority.

Not all of my encounters with closed-minded, status-quo scientists are confrontational. And even interviews like this one with Dr. Graziano usually end on a positive note. In this case, he seemed sincere when he thanked me for a "lively discussion." But for others, fielding tough questions about science's absurd assumptions can be overwhelming.

Dr. Patricia Churchland

I feel obligated to offer some kind of warning before sharing this next interview transcript. Although Dr. Patricia Churchland is a recognized expert in consciousness research and is regularly called upon to give television and radio interviews, her performance on *Skeptiko* was unlike anything I've ever experienced. I can still remember finishing the interview and wandering into the kitchen for a debriefing with my wife. Fortunately for me, Joni is a licensed psychologist so the process of talking through the strange interview served as sanctioned therapy.

Before you read the transcript there are a couple of concepts you need to be familiar with. The first is "emergence." Dr. Bernardo Kastrup does an excellent job of explaining how emergence relates to consciousness in Chapter 12, but for now what you need to know is that some consciousness researchers like Dr. Churchland have convinced themselves that consciousness must be some kind of "emergent property" of the brain. To them, consciousness rises out of the electrochemical soup inside our head like sand dunes emerge in the desert. It wouldn't be such a terrible theory if it offered any explanation for how, when, or why consciousness exists. But since it doesn't, it's little more than a passing-the-buck exercise by scientists determined to prop up the "consciousness is an illusion" theory.

The other idea requiring explanation is, "consciousness is an epiphenomenon of the brain." Simply stated this means brains "make" consciousness. Most of the "consciousness is an illusion" folks wind up coming around to this idea. They often engage in a lot of arm-waving and over-intellectualizing in order to avoid saying it, but it's what they mean.

So, if you don't like confrontational interviews, you might want to skip over this one with well-known consciousness expert and University of California at San Diego professor Dr. Patricia Churchland (*Skeptiko* #237):

Me: *Yeah, but aren't you trying to move away from the "consciousness is an illusion thing" without jumping all the way to the other side where the physicists are taking us? Physicists are saying consciousness is somehow fundamental.*

If we break down this debate on the nature of consciousness, we have these two camps... One is this very materialistic view, like I think you started out, but then I don't know if you really were holding to that—that consciousness is purely a result of an epiphenomenon of the brain...

Dr. Patricia Churchland: *No, it's not an epiphenomenon. It is an actual phenomenon in the physical brain. It's one of the things that the physical brain does in just the way that your brain stores* memories. Some of those memories change over time as a result of changes in the physical brain. We know, for example, that people who have Alzheimer's, because they have lost many neurons in their brains, no longer have the capacity to remember certain things. Memory is a real function of the physical brain and so is consciousness. It's not an illusion; it's the real deal.

Me: But what is [consciousness]? Don't we need to nail it down a little bit more than that? You're saying it's an emergent property of the brain. Isn't that kind of passing the buck a little bit?

Here's the other possible explanation. Consciousness is somehow fundamental and the brain is somehow interacting with [it]. Somehow the field of consciousness is out there, and [our] brain is somehow interacting with it. But that's not to confuse it with being purely a result of brain activity. I mean, that is a completely different theory, right?

Dr. Churchland: It's a theory for which there is essentially no evidence. One of the problems with that approach is that we can't understand why taking a drug, for example, should change your consciousness if consciousness is not part of the physical brain because we know that the drug changes the physical brain and that consciousness is somehow completely independent of that because it's a fundamental feature of the universe.

Me: It doesn't have to be completely independent. Obviously there's some relationship, a close relationship...

Dr. Churchland: Ah, okay. What always puzzled Descartes is if there is an independent non-physical soul, how does it interact with the physical brain? The problem with dualism is that nobody has ever been able to address that in a meaningful, testable way.

Me: Yeah, but I think we also have problems with the idea that it's the emergent property of the brain thing that we're moving to, right?

...I don't really know how that answers the big questions of what the nature of consciousness is other than just to repeat that consciousness is something that brains do. That doesn't tell us much. How does it begin? When does it end? What's necessary and sufficient to cause consciousness? These are all questions that are unanswered by what you're saying.

Dr. Churchland: *Well, neuroscience hasn't got all the answers yet.*

Me: *But that's just passing the buck. We don't have the answers? Those are fundamental questions. If we don't have the answers then we don't have a theory of what consciousness is, right?*

Dr. Churchland: *That's what your view seems to be, all right.*

Me: *I'm just saying these are the basics. When does consciousness begin? When does it end? What is necessary and sufficient to create consciousness? If we can't answer those [questions] then what do we really have? What can we really say about consciousness?*

Dr. Churchland: *Well, I guess we can't say anything.*

Me: *Okay. I think we can say some things. Let me ask you this—I didn't mean to throw you completely off. Do you want to get back to talking about your book?*

Dr. Churchland: *No, not really.*

She was clearly getting upset, but I persisted:

Me: *Okay. Let me ask you this, Dr. Churchland. Do you think consciousness can do work? Is that an important element to understanding what consciousness is? I mean the [present] theory assumes consciousness can't really do anything. Have we come around? I think there's research suggesting maybe consciousness can be focused, and can direct and affect neuroplasticity. What's your feeling about that?*

Dr. Churchland: *Well, you seem to think that it does work so why don't we just go with that?*

Me: *So you agree. It does do work?*

Dr. Churchland: *I have no opinion on the matter.*

Me: *Okay. What do you think about near-death experience? You write quite a bit about that in your book, [so] what is your general* take on near-death experience?

Dr. Churchland: Well, I'm not sure that it really matters, does it? What does it matter for?

Me: I think a lot of folks look at near-death experience as highly suggestive of consciousness somehow, in some way we don't understand, surviving biological death, which would certainly falsify the idea that consciousness ends at death. I mean, that would falsify that, right?

Dr. Churchland: Oh, I'm sorry. My dog just came in. No, no, don't do that. No, no, no, no. Forgive me, I'm sorry. Okay. So yeah, okay, I guess I've never actually had a near-death experience. Have you?

Me: No.

Dr. Churchland: Oh, okay.

Me: But you write quite a bit about it in your book.

Dr. Churchland: So why do you want me to talk about it?

Me: Well, I guess one of the things I did want to ask you is in your book you ask the question "Is there a neurobiological explanation for near-death experience?" Then you cite NDE researcher and a former guest on this show as answering that question with, "yes." You say that Dr. Pim van Lommel believes the answer is "yes." Is that your understanding of his research?

Dr. Churchland: Well, I think there's certainly quite a bit of evidence that at least some near-death experiences have a neurobiological basis. Of course, we can't be sure about all of them. Maybe you had one that doesn't have a neurobiological basis. I wouldn't really know, would I?

Me: Well specifically, Dr. Churchland, you cite in your book that Dr. Pim van Lommel holds that opinion. That's clearly not the case. I mean, he's written...

Dr. Churchland: Has he? Uh-huh [Yes].

Me: Right. Do you want me to read to you what he's written? He's written that "The study of patients with near-death experience..."

and this is from The Lancet *paper that you're citing, "clearly shows us that..."*

I was mid-sentence when she hung up. This was a first. Of all the interviews I had done, no one ever hung up on me. I immediately sent Churchland an email. I encouraged her to complete the interview and asked her to consider how she was going to be perceived if she ran from this debate. To my surprise, she agreed, and within minutes we were once again on a Skype call:

Me: *Well, let's try and finish. I think it was getting a little bit testy there. Tell me what the rub is here. I'm telling you that you totally distorted Van Lommel's quote. It's right there in your book. I can give you the exact page. It's on page 71. You say, "There's a neurological explanation for NDEs." And then you cite Pim van Lommel as a NDE researcher who says there is. He clearly doesn't; he says the exact opposite.*

Dr. Churchland: *[Silence]*

The second call didn't go any better than the first. In some ways it was worse because Dr. Churchland knew what I was going to ask. After a few tense exchanges she hung up on me again, and I was sure we'd reached the end. Her claims of technical problems were hard to take seriously. If you listen to the recording, you can hear background noise at the same time she claimed to be having problems with her microphone. Nonetheless, I sent her one more email after the second hang-up. I offered to call her on a landline. I offered to conduct the interview by email. But she always seemed to have a reason why she couldn't complete the interview. I've never heard from Dr. Churchland since.

Some *Skeptiko* listeners thought Dr. Churchland looked childish. A few thought I shouldn't publish the interview. But I did publish it because I believe trusted academics like Dr. Patricia Churchland have a responsibility to enter the court of public opinion and share their tax-dollar-funded opinions, no matter how it makes them look.

With all this talk of consciousness, brains, and neuroscience you might think the problems of status quo science are limited to a few narrow areas of research. But as you're about to see, the problem science has with consciousness has an impact on every aspect of science-as-we-know-it.

CHAPTER 2:
WRONG ABOUT QUANTUM PHYSICS...IF A TREE FALLS...

Quantum mechanics is stunningly successful. Not a single prediction of the theory has ever been wrong... However, quantum mechanics also displays an enigma. It tells us that physical reality is created by observation, and it has "spooky actions" instantaneously influencing events far from each other—without any physical force involved. Seen from a human perspective, quantum mechanics has physics encountering consciousness.[8] — Physicist Bruce Rosenblum

    A long time ago philosophers asked the question, "If a tree falls in the forest and no one is there to hear it, does it make any sound?" This is a question about the nature of consciousness. It asks the same question we considered in the last chapter—can consciousness extend beyond the physical brain?
    This may sound like a strange question, but it turns out to be related to an enigma quantum physics has wrestled with for the last 100 years—the "observer effect." I won't bore you with the details, but the bottom line is this: quantum physics is not kind to the idea, "I am in here, and the world is out there."

Dr. Mario Beauregard

The leap from quantum physics to the study of consciousness is not as big as it might seem. In 2012 I had the opportunity to talk with Dr. Mario Beauregard (*Skeptiko* #195), a neuroscientist from the Neuroscience Research Center at the University of Montreal, about his book *Brain Wars*.[9] During the interview Dr. Beauregard explained the current paradigm that dominates science-as-we-know-it:

Dr. Mario Beauregard: *The starting point is what we call the "modern scientific worldview." It's a worldview based on classical physics and a number of fundamental assumptions like materialism, determinism, reductionism. So applied to mind and brain it means that, for instance, everything in the universe is only matter and energy, and the mind can be reduced strictly to electrical and chemical processes in the brain...*

Me: *Dr. Beauregard, is that really where science is at?*

...Do most people really believe any of that? And to what extent is that really the mainstream view within science?

Dr. Beauregard: *Well, it's still the mainstream view in the biomedical field. It's also the mainstream within philosophy. However, in physics everything changed 100 years ago when they had their own revolution away from classical physics to quantum physics and the de-materialized world. So the founding fathers of quantum physics realized that the universe is not constituted of tiny physical particles like billiard balls, for instance.*

They also realized that there's a very important mental component in the universe, and this relates to the so-called "measurement problem" or "the observer effect." The observer effect showed physicists that, to a certain extent, their observation of the experiment was affecting the outcome of the experiment at the micro-physical level. And so now it's recognized in quantum physics that you cannot explain the universe or understand the universe without making reference to what we call mind and consciousness.

But in the bio-medical field, including neuroscience, it's a differ-

ent ballgame. *Most scientists in my field of research, for instance, are not aware of this. They don't know quantum physics, and they're not aware that the fundamental issues in physics went through this revolution 100 years ago. Most scientists and philosophers are either not aware of that or they don't want to accept it.*

Later:

Me: *You do a nice job of pointing out how the existing scientific paradigm is frayed at the edges. I like the way you tie together a lot of science we're familiar with and show us how it violates scientific materialism; for example the placebo effect, and also hypnosis... Tell us how those violate the very strict understanding of reductionist materialism.*

Dr. Beauregard: *For a number of materialist thinkers and scientists, the mind is totally powerless. It cannot exert any power on what's going on at the brain level and anywhere in the body.*

Me: *Can you break that down and explain why that would have to be true if you're a strict scientific materialist?*

Dr. Beauregard: *There are a number of materialist positions related to this. One is called epiphenomenalism. These proponents recognize that my mental processes do exist but believe they are powerless. They cannot exert any influence.*

That's one position, but you have other positions like eliminative reductionism where you're trying to eliminate all mental processes. So, Philosophers Daniel Dennett or Churchland will say that consciousness and all the other mental processes are simply illusions; that the only thing that exists is electrochemical activity in the brain.

If this is true, then of course you cannot influence the activity of your brain by your beliefs or your expectations. But that is exactly what the placebo demonstrates. The placebo effect shows that your beliefs and expectations can significantly alter what's happening in your brain and in the physiological systems connected to your brain.

For example, in the last decade there have been several brain and aging studies focusing on the placebo effect. In one case there was a very interesting study done at the University of British Columbia. They did a study to measure the impact of the placebo treatment on people suffering from a severe form of Parkinson's disease. In Parkinson's disease there's a great level of destruction of the nerve cells, specifically the neuron producing a chemical messenger we call "dopamine." Dopamine is the key chemical messenger in motor function, but it's also involved in many other activities.

In this case the patients had a level of nerve damage of about 70% to 80%. So, the level of destruction of the nerve cells that produce dopamine was quite high. And, of course, the patients were severely impaired from a clinical point of view. They had trouble moving. They were experiencing a lot of tremors. So the neurologists doing the study presented them a fake treatment. It was only distilled water, but they told the patients that this was potentially a revolutionary treatment for Parkinson's disease.

A few minutes after the injection, they scanned them with a technology called positive emission tomography. They were interested in measuring the activity of dopamine in the brain. Very rapidly, those patients who believed in the bogus treatment started to produce and release dopamine into their brains in an amount comparable to that seen in young, healthy people! And clinically, they started to improve. They had less tremors, more strength, and were more optimistic, at least for a certain period of time.

This is a very nice illustration of the power of what we call "mind." By mind, I mean all our mental activity and mental events. In this specific case, the patients' improvement was related simply to the beliefs and expectations that the patients had regarding the fake treatment. They believed they would get better and they did.

Later, Dr. Beauregard discussed how hypnosis also points to the existence of a "mind" that is more than just our brain:

Dr. Beauregard: *Another interesting example is the use of hypnosis. Hypnosis is based on suggestions coming from a hypnotist and the subject accepting [those] suggestions. Now the experts in* this field of research consider all forms of hypnosis are in reality self-hypnosis. That means that if the subject does not want to accept the suggestions from the hypnotist they won't work.

So again, during the last decade there have been a number of brain-imaging studies measuring whether there's something happening at the brain level during hypnosis. This is partially in response to skeptics who have been arguing for a long time that hypnosis is simply wishful thinking. They say it's only based on social compliance from the subjects who want to please the hypnotist.

Several research teams have attempted to research this question using brain imaging. For instance, there was a very interesting study done at Harvard. The researchers were looking for the neural correlates of color vision. Correlates are physical processes in the brain, and whether electrical or chemical, they are related to a specific mental activity. For instance, the perception of color. So in this study they scanned the subjects, highly hypnotizable subjects, which means that these people had the capacity to easily enter a deep trance state. And so these subjects were able to imagine that they were seeing different types of patterned colors in accordance with the suggestions of the researchers doing the study. It was very interesting because when they were presenting subjects with images that were colored, the researchers were telling the subjects these were only gray. And when they did this, the brain regions associated with the processing of color were not activated.

The reverse effect was also measured in this study. That is, when there was no color in the image, the researchers told subjects there was. In this case the scans showed there was still activation in the regions of the brain processing colors. So again, this is a very nice illustration that what's going on at the mind level can exert a great influence over what's happening in terms of brain activity.

When we *follow the data, wherever it leads,* we quickly run into a lot of experimental evidence that contradicts this *you-are-a-biological-robot* assumption science has been stuck on. Then again, you always knew this! This biological robot stuff is an absurd idea that goes against everything we know and experience. People don't really believe their life is meaningless. We've just been conditioned to respect the authority of those over-credentialed scientists. We've been conditioned to follow *the creed*.

CHAPTER 3:
WRONG ABOUT SCIENCE...
AND ITS DOPEY CREED...

Scientists are specialists... [who] have never even grasped the crucial distinction between wisdom and speculative hypothesis based on experiment...When they venture outside the confines of their specialties and try to justify their theories and effects in terms of value or to assess their metaphysical or human significance, they produce only nonsense... Judged by the normal standards of metaphysical or philosophical discourse, scientists—but for the most rare exception—display a total lack of competence in this realm.[10]

— Philosopher Philip Sherrard

My upbringing in the Greek Orthodox Church was a strange experience. I remember sitting in an incense-filled cathedral watching grown men march in ceremonial robes, carrying gold-plated icons, all the while murmuring prayers in a language I didn't understand. I didn't get it. I got the fear part—it scared the heck out of me—but the awe and reverence it was meant to inspire never stuck. Even as a kid I could see the hypocrisy of following a creed you don't believe in. Science-as-we-know-it finds itself in a similar situation. It is married to a creed few really believe in, but everyone is afraid to let go of it.

In 2008, I had a chance to interview a delightfully clear-thinking psychologist named Dr. Charles Tart (*Skeptiko* #6) about his book, *The End of Materialism*.[11] In it, Tart explains how science's inability to account for consciousness has led to the adoption of an unspoken creed. I've adapted Dr. Tart's ideas into something I call, *The Dopey Science Creed*:

1. I maintain that my life has no purpose and no meaning. The same is true for the entire universe. There is no purpose to anything.

2. I affirm that my morals come from my genes and my conditioning, not from decisions I make. Free will is an illusion. My personal identity is an illusion.

3. There are no "good" deeds, or "good people." There is no "bad," "evil," or "wrong" either.

4. Every report of encounters with spirits, angels, ghosts, and supernatural beings is bunk. The credibility or number of witnesses doesn't matter—it's all bunk.

5. I am my physical brain and nothing more. The death of my body is the death of me.

Most of us are rightfully put off by at least some parts of this creed. Even the most scientifically-minded parent can't look their son or daughter in the eye and tell them they are a meaningless illusion. But while you might feel uneasy with the Dopey Science Creed, science-as-we-know-it doesn't allow for wiggle room on these points.

Take the first element of the creed, "Life has no purpose or meaning." Although the claim sounds harsh, it's not as though science is trying to tell us we're meaningless peons. The problem is science has cornered itself into a position that allows for nothing else. If life has meaning, then what is the meaning of life? Who determines life's meaning? How do we measure it?

To suggest anything in the universe has "meaning," other than illusions of meaning we ascribe with our robot brain, is blasphemous to science because it's an admission there's more to life than science is able to measure.

And then there's the question of bodily death. Even though science-as-we-know-it is totally incapable of explaining how that little voice inside your head (i.e. consciousness) got started, they're sure about one thing—when you die it's gone for good. Again, it can't be otherwise for science because you, they claim, are your brain. The end of your brain has to mean the end of you.

But this assumption creates another "meaning" problem for science. If your physical death is the end of you, then what possible meaning could there be to your life. Why love? Why care? Why do anything? If everything is going to disappear when you die, then why live? It's an impossible question. And, if you think the future of your children can get you out of this mess, they can't. Why pass along your genes if only to perpetuate an absurd illusion of existence? What possible purpose could there be in this meaningless universe science has postulated? Death, it would appear, is the great teacher science chooses to ignore.

Matt Dillahunty

The Dopey Science Creed has become inseparable from mainstream science. It's the framework for everything science could know about life. It's also the lens through which atheists see the world. Modern-day atheists sometimes referred to as "New Atheists," use this brand of science as the basis for their belief system. On *Skeptiko*, I've managed to incorporate elements of the Dopey Science Creed into a number of my dialogues, but never more directly than during this short email interview with public speaker, internet personality, and former president of the Atheist Community of Austin, Texas, Matt Dillahunty.

In response to point #1 of the Dopey Science Creed (There is no purpose to anything), he replied:

Matt Dillahunty: *I see no reason to accept assertions that there* is some externally imposed, agent-guided purpose to life. My life has plenty of meaning and purpose—we imbue things with meaning and purpose. We even specifically create things with an intended purpose in mind—so it's hyperbolic, at best and simply false, at worst to claim, "There is no purpose to anything." It's sloppy wording that attempts to straw man the actual position, which is: The assertion that life has an externally-imposed, agent-guided purpose has not met its burden of proof.

Me: Ok, but that is logically inconsistent... that's why [Tufts University Professor, and well known thinker in the New Atheism movement] Daniel Dennett asserts that "consciousness is an illusion." Either the meaning and purpose you give life is an illusion, or it's not.

Moreover, this is nonsense because no one lives their life like this. No one thinks the love they have for their family, their children, their closest friends is all a trick of the mind. In fact, anyone who really believes this would have no reason to live... why perpetuate an absurd illusion?

In response to point #2 (There is no "right" or "wrong"), he replied:

Dillahunty: This is a mess of combining issues. Free will may be an illusion, depending on how it is defined—but that might be irrelevant to issues of morality. I've lectured on the superiority of secular morality many times (there are several different talks online) and I not only advocate for right and wrong (not as extant things but as non-subjective values) but also moral absolutes (with the caveat that each situation is considered independently) and I make no appeal to genes or conditioning or claim that there is no "me." I have consistently and repeatedly rejected moral relativism...as has Sam Harris.

Me: Well, you might have to lecture us one more time because "secular morality" doesn't make any sense in a meaningless universe. The fact that you and others have bought into [neuroscientist

Wrong about Science... And Its Dopey Creed...

and best-selling author on atheism] Sam Harris's sleight of hand doesn't make it any more defensible.

In response to point #3 (There are no "good/bad people"), he replied:

Dillahunty: *This is another mess of combining issues. What does "good people" mean? If I had to guess, I think I'll accept the notion that people, generally speaking, aren't wholly good or bad...or intrinsically good or bad, but instead there are people who do good things and people who do bad things. There are certainly people for whom the quality or quantity of good or bad things they've done is most easily summarized by calling them "good" or "bad"...but that's a shortcut generalization that is more colloquial than philosophical.*

There are most definitely people who have been labeled mystics, sages, prophets and saints. Whether or not the abilities attributed to them are real, or not, is a separate question. But I don't (and most skeptical atheists I've met wouldn't) assert that these claims are false...merely that these claims haven't met their burden of proof and can't rationally be considered to be true.

Me: *Again we're hashing over the same issue... if there is no "meaning" then there's no objective reality to such ideas as "good" and "bad"... forget about your notions of "wholly good or bad" there's not even "somewhat good or bad."*

Your buddy Sam Harris (Skeptiko #189 and #192) tries to get around this problem by appealing to neuroscience and claiming that we know enough about the brain and its relationship to the physical world so as to allow a new "scientific mapping of good and bad." This is silly on a number of levels, but most importantly for our discussion it is rigidly wed to the idea that mind equals brain. If there's any case where mind is not equal to brain, then his idea crumbles.

In response to point #4 (Encounters with spirits, angles, ghosts and supernatural beings are bunk), he replied:

Dillahunty: *The response here is the same as for the last one. I* don't believe the claims are true. That doesn't mean that I believe the claims are false. The truth of a claim isn't in any way impacted by the number of people who accept it, nor by their apparent credibility, nor by the sincerity or degree of their conviction. But pointing out that a claim hasn't met its burden of proof and cannot rationally be considered "true" is NOT the same as claiming that the claim is false.

You've constructed a straw man that equates "I don't believe this is true" with "I believe this is false"...which is simply false.

Me: Well, thank goodness for scientific methods, and peer review and all that stuff or we'd just have an endless chain of "straw man" charges by those who claim to be standard bearers for the "burden of proof."

In response to point #5 (I am my brain. The death of my body is the death of me), he replied:

Dillahunty: In many ways, "I am my brain" is true. Is it exhaustively true, in all contexts? Nope. But when talking about who I am, those aspects produced by my brain certainly have primacy. As far as we can tell, when my brain dies, I cease to exist. There's no demonstrated mechanism for consciousness to persist beyond death. There are a lot of complicated discussions to be had about identity and self... but you gloss over that in the creed for a simple assertion followed by another exaggerated straw man. I do not assert that the claimed encounters people have had are ALL an illusion. I don't propose an explanation for ALL of them. I don't necessarily propose an explanation for ANY of them (though we can be reasonably confident that delusions do occur).

My position isn't "All encounters with those who have died are an illusion." It's "No encounters with those who have died have been confirmed to actually be as claimed."

There's a big difference there—and it's one you've repeatedly misrepresented in your creed.

But when talking about who I am, those aspects produced by my

Wrong about Science... And Its Dopey Creed...

brain certainly have primacy. As far as we can tell, when my brain dies, I cease to exist. There's no demonstrated mechanism for consciousness to persist beyond death.

Me: Great....now we're getting somewhere....tell me when mind-equals-brain is not true?

Also, you say, "There's no demonstrated mechanism for consciousness to persist beyond death." What is the "demonstrated mechanism" for consciousness to exist before death?

Ok, Matt, this has been a great first round of dialogue...hope we can keep it going as I think we're quickly moving into the issues we care most about on Skeptiko—what does consciousness science tells us about who we are?

I had planned on a lengthy point-by-point debate with Dillahunty when I created a thread on the *Skeptiko* forum for us. But Dillahunty, like so many of the atheists I've encountered, seemed to disappear when the conversation veered away from the talking points atheists usually lean on.

While Dillahunty isn't a heavyweight scientist like some of the guests that have appeared on *Skeptiko*, he performed no worse when handling these tough questions in our email exchange. But he didn't do very well either. Of course, it's all not his fault. The materialist, science-as-we-know-it creed he's following is contradictory and absurd. If life is meaningless, then why assert anything? Why defend anything? Why do anything?

Conversely, if you think your life does have meaning, where does that meaning come from? And, before you answer, remember you can't say "you" give your life meaning because "you," according to science-as-we-know-it, are an illusion.

The Dopey Science Creed is "dopey" because no one lives his or her life thinking it's an illusion. The absurdity is self-evident. But it's also "dopey" because its underlying assumption—that your minute-by-minute experience is nothing more than electrochemical processes of your brain—is testable. It is falsifiable by experimentation.

We'll explore some of those experiments in the next few chapters.

CHAPTER 4: WRONG ABOUT NEAR DEATH EXPERIENCE...

In science... false theories arise, receive support, and sometimes thrive, not so much because of false data or invalid reasoning, but because scientists take for granted assumptions that are actually untrue, and they neglect to re-examine all assumptions whenever significant new information becomes available.[12]
— Astronomer Tom Van Flandern

I've never had a near-death experience. No one in my family has had a near-death experience. And before I started *Skeptiko*, I had never heard of near-death-experience science. This might explain why I felt a need to do more than 40 shows on the topic; I had a lot to learn.

The Near Death Experience (NDE) phenomenon is an extraordinary medical and scientific mystery that shatters our science-as-we-know-it understanding of who we are. Nothing is more threatening to science than evidence of consciousness surviving death.

When Dr. Raymond Moody, M.D., a two-time *Skeptiko* guest (*Skeptiko* #21 & #174), published *Life After Life* in 1975, he opened the door to scientific research into what lies beyond bodily death.

But since then only a handful of researchers have been brave enough to follow his lead. Over the last seven years I've had the chance to interview many of the world's leading NDE researchers, including cardiologist Dr. Pim van Lommel (*Skeptiko* # 119), oncologist Dr. Jeffrey Long (*Skeptiko* #99 & 118), medical researcher Dr. Penny Sartori (*Skeptiko* # 90), cardiac arrest and brain resuscitation expert Dr. Sam Parnia (*Skeptiko* # 116), professor of counseling Dr. Jan Holden (*Skeptiko* #164), and neuropsychiatrist Dr. Peter Fenwick (*Skeptiko* # 64).

One thing all these researchers have in common is their scientific approach to the topic. None were "believers" before they began their research. And when you listen to the interviews I did with them, you'll discover that a couple of them were firmly in the "non-believer" camp.

I've also interviewed near-death experiencers who had dramatic spiritual transformations leading them to do scholarly research of NDEs. Former Harvard neurosurgeon Dr. Eben Alexander (*Skeptiko* #220) and Dr. Suzanne Gordon (*Skeptiko* #214) come to mind. But whether they were experiencers beforehand or not, all the NDE researchers I've encountered have come to the same conclusion: *consciousness survives death*. And while this conclusion is sometimes presented in language designed to lessen the shock to their colleagues, they all say it. They may say the evidence "seems to suggest" or "defies our current neurological models," but when you talk to them in conversation you find that every serious NDE researcher has become—for lack of a better word—a "believer."

Scientific Consensus Is Not Enough for True Non-Believers

Naturally, you might wonder how this overwhelming consensus among respected researchers has impacted mainstream science. I mean, science is driven by evidence, right? Publish a hundred unchallenged peer-reviewed papers and opinions change, right? Nope.

In fact, while I was interviewing some of the top NDE researchers, I was also talking with some of the most well-known, and most vocal, NDE critics, including neurologists Dr. Kevin Nelson (*Skeptiko* #93) and Dr. Steven Novella (*Skeptiko* #16, #44, & #105), philosopher Dr. Massimo Pigliucci (*Skeptiko* #107), science writer Dr. Michael Shermer (*Skeptiko* # 3), anesthesiologist Dr. Gerald M. Woerlee (*Skeptiko* # 98), and parapsychologists Dr. Caroline Watt (*Skeptiko* #165) and Dr. Susan Blackmore (*Skeptiko* # 5 & #114).[13]

What I discovered from the NDE critics surprised me. First, most have never actually researched near-death experiences. This is true of Novella, Shermer, and Pigliucci. The sum total of their scholarly publications in this field is zero.

Second, even those who have published on the subject, like Dr. Caroline Watt, Dr. Susan Blackmore, and G. M. Woerlee, haven't actually worked with near-death experiencers. Although these "experts" are frequently cited by the mainstream science media, their "research" consists of little more than academic-sounding book reports.

When you push this point, NDE critics will usually admit their lack of depth in the field, but most are never asked. When I interviewed Dr. Watt from the University of Edinburgh about her often-cited, strongly assertive paper explaining away NDE science, she admitted:

Dr. Caroline Watt: *...this is actually not my area of specialty. It's probably my one foray into near-death experiences, and I probably won't be publishing on it again.*

Similarly, despite her ongoing status as an authority on NDE research, Dr. Susan Blackmore admitted to me in 2010 (*Skeptiko* #210) that she wasn't an expert and hadn't remained current in the field for the last 20 years:

Dr. Susan Blackmore: *It's absolutely true. I haven't written about this subject for a long time, and I haven't kept up with all the literature, either.*

When pressed further about some of the negative comments

she had made about the work of NDE researcher Dr. Jeffrey Long, Blackmore admitted:

Dr. Blackmore: *I gave up all of this stuff so many years ago. If you are a researcher in the field, it behooves you to read as much as you can of the best work, because otherwise you can't be a researcher in the field. I'm not a researcher in the field. I have not been for a long time.*

Soon after my interview with her in 2010, Dr. Blackmore gave a public presentation on her pessimistic view of NDE research. I wonder if she mentioned her lack of expertise to her audience.

Shoddy Research Exposed

If the evidence in favor of the NDE hypothesis—that consciousness, in some way we don't fully understand, continues after brain death—is so strong, why are there so many scientific sounding papers refuting it? This was the question I sought to answer when I interviewed Dr. Caroline Watt, about her paper with Dean Mobbs, "There Is Nothing Paranormal about Near-Death Experiences: How Neuroscience Can Explain Seeing Bright Lights, Meeting the Dead, or Being Convinced You Are One of Them."[14] I had doubts about the paper, so I wanted to hear from the author herself. I wanted to know how this paper with no original research had passed peer review and gained so much media attention.

But before interviewing Dr. Watt, I wanted to talk to a qualified NDE researcher who could help me prepare. I chose NDE researcher and co-editor of the scholarly *Handbook of Near-Death Experiences*, Dr. Jan Holden of the University of North Texas. During the interview I asked what she and other NDE researchers thought of Dr. Watt's paper:

Dr. Jan Holden : *I didn't see any original research. What I saw was a compilation of theories and results that have been published for quite some time, and have been answered in* The Handbook of Near-Death Experiences.[15] *What I noticed about this article is that* *it's citing a lot of old sources that have already been responded to.*

Me: *Let's get to the meat of their paper. I'll give you this quote: "Contrary to popular belief, research suggests that there is nothing paranormal about these experiences. Instead, near-death experiences are the manifestation of normal brain function gone awry." I know from your continuing education course on near-death-experience science that there are at least 10 prospective NDE studies with in-hospital patients. I don't think one of them would support this conclusion.*

Dr. Holden: *The research material that's out there actually supports a different conclusion. To quote my colleague Bruce Greyson,[16] "If you ignore everything paranormal about NDEs, then it's easy to conclude that there is nothing paranormal about them." And that's what they have done.*

During my subsequent interview with Dr. Watt, I asked her to explain how this peer-reviewed journal article, with so many glaring mistakes, misrepresentations (including a significant misquote of one of its primary sources), had unashamedly ignored the primary work of the field:

Me: *I'm saying that what gets picked up and perpetuated through the science media is reflective of the current position, even if that position isn't supported by the best data. I'm saying your paper got traction even though there's not a lot behind it. I'm saying you cited references incorrectly. And you referenced skeptics like Dr. Susan Blackmore, who admit to not being current in the field.*

Dr. Watt: *As I said, it was intended to be a provocative piece. It's not claiming to be balanced.*[17]

An Unfair Fight

It's stunning to hear a scientist admit their work was *not* intended to be balanced, but the NDE phenomenon seems to cause otherwise rational, logical scientists to act strangely. In 2010, I interviewed Dutch anesthesiologist G. M. Woerlee. This short segment shows how strange it can get:

Me: *But Dr. Woerlee, you are saying that any account that describes hearing or seeing without the brain being present should be dismissed. How can we ever get out of circular logic if we start with the assumption that there isn't any possible way that the out-of-body experiencer can really be seeing and hearing?*

Dr. G.M. Woerlee: *There is one very simple way, and that is that the out-of-body experiencer is actually seeing and hearing with their body. They are hearing with their ears, and that means they can hear the sound. And they see with their eyes in some cases, because their eyes are open, so they can actually see people around them. When you close your eyes, you can actually visualize in your mind's eye, as it were, quite a lot of what is happening around you.*

Me: *But again, that doesn't fit the data. It's an explanation, but when we try to match it up with the data, it doesn't fit. What I think the out-of-body experiencers would immediately point out is that they're often bringing in information they couldn't possibly see within their immediate vicinity. They are able to sometimes hear things and see things down the hall of the hospital, or on a different floor, or miles away.*[18]

My interview with Dr. Woerlee created an opportunity for me to invite back to *Skeptiko* Dr. Jeffrey Long (*Skeptiko* #99), a radiation oncologist and NDE researcher. Dr. Long had appeared on *Skeptiko* following the release of his best-selling book, *Evidence of the Afterlife: The Science of Near-Death Experiences*.[19] This book is based on the most comprehensive study of NDEs ever compiled.

I admired Dr. Long's in-depth research of NDEs and his ability to communicate his findings so clearly and eloquently. I also admired his willingness to brave the academic and cultural taboos his colleagues shy away from. In my interview with Dr. Long, I discussed the methodology he used for his large-scale survey carried out through a specially designed website (www.nderf.org):

Me: *...I was watching the interview you did with Bill O'Reilly* on Fox. *I thought that was a great example of where a guy, Bill O'Reilly in this case, becomes completely sidetracked with this "anecdote" thing. [As in:] "These are just anecdotes and we can't really trust them because they're anecdotes."*

So I thought you should talk a little bit about survey work in medical research and that how when you ask one person about their NDE, you may have an anecdote, but when you compile a 150 question survey of 600 people, you have something much more than that.

Dr. Jeffrey Long: You almost just answered it right there. No doubt about it. When you have an anecdote, it simply means one or a few case studies. You have to be very hesitant to draw conclusions about small numbers. It would be like testing a new drug on three patients and trying to find sweeping conclusions about its effectiveness. You really can't. You need much more data than that.

That's why the Near Death Experience Research Foundation (NDERF) study that I presented in [Evidence of the Afterlife] is certainly vastly beyond anecdotal evidence. We actually studied 1,300 near-death experiences. It's certainly not just a limited number of case reports. And you're right, our modern questionnaire is over 150 questions, so no doubt we have the depth of analysis, as well. And most of the research that's published in the book was based on surveying over 600 near-death experiencers that filled out the most recent version of the questionnaire.

Let me start out with sort of a basic scientific overview, and that is what's real is consistently observed. So we've observed evidence of the afterlife and near-death experience, not only in the vast number of near-death experiences studied in tremendous depth in my own study, but all my major findings are corroborated by scores of prior scholarly studies. We're way beyond what could reasonably be called anecdotal. We're really in very hardcore, evidence-based [findings when it comes to my work] and the work of many others.

Me: *One other point that came up, and this was in the interview* that I had with Dr. Woerlee, [first of all,] he acknowledged that your research is solid; that your database is solid, which is good to hear. But then he had one point about the survey methodology that I'd like to talk about. He suggested that perhaps your survey is biased in favor of people who have a certain kind of NDE.

Now he may have articulated it a little bit differently, but can you talk about any concerns that someone might have that your survey is biased or that you're just pulling accounts of a certain kind?

Dr. Jeffrey Long: *His actual quote, as I recall, was referencing the fact that if people had basically non-near-death experiences, or very brief experiences, that they wouldn't necessarily be in the research. That's not true at all. What we did with our NDERF study is we studied every single person who had a near-death experience. In other words, they nearly died and they had an experience.*

In addition to that, we used the most validated research tool in near-death experience research, and that's called the NDE scale. So we analyzed every single person that had such an account. In fact, we post every single near-death experience on the website [as long as they] give us permission, which is over 95%. So we not only have a very valid, comprehensive look at near-death experience because of the numbers, but in addition to that we share that with the world, so everybody else can see the data set that we're seeing, too.

We then went on to examine some of the specific criticisms that Dr. Woerlee had leveled against the nine lines of evidence presented in Dr. Long's book. The first argument we examined was Woerlee's claim that near-death experiences while under general anesthesia could be explained as "anesthesia awareness."

Dr. Jeffrey Long: *... I talk directly about this so-called anesthetic-awareness that Dr. Woerlee discusses.*

As an overview, let me say that these anesthetic-awareness experiences are so very, very rare that I hope this never dissuades anybody from having medically appropriate general anesthesia. Please don't let any of the discussion here be an issue in preventing appropriate medical care. As I say in the book, and this is a direct quote, "Rather than the type of coherent NDEs you read here, anesthetic-awareness results in a totally different experience." And I provide a number of references on that, by the way, for interested listeners.

I go on to say, "Those who experience anesthetic-awareness often report very unpleasant, painful and frightening experiences. Unlike NDEs which are predominately visual experiences, this partial awakening during anesthesia more often involves brief and fragmented experiences that may involve hearing but usually not vision." Again, I emphasize that anesthetic-awareness is very rare under anesthesia.

By the way, I'm not aware of any near-death experiences that occurred under general anesthesia on the NDERF website that described the typical content of anesthetic-awareness experiences. Dr. Woerlee brings up a few anecdotal discussions about anesthetic-awareness, but I have a number of [other] references [in the book]. These are the scholarly people that have actually studied a number of anesthetic-awareness experiences and published them in peer-reviewed journals in the past. That's my source.

...These are completely different experiences, anesthetic-awareness and near-death experiences. I don't think Dr. Woerlee quite got that point how clear that was; how crystal clear the distinction between those two types of experiences is.

Me: ... How do you counter these arguments that don't really make sense, but [are paid attention to] because they come from people who should know better, and because they're said with such conviction [and repeated] over and over again? Even though Dr. Woerlee was picking out anecdotes he thought were appropriate, they still didn't even match the data... And yet [I was] never able to get there in my discussion with him. How do you engage? How do you carry out a scholarly or an intellectual discussion along those lines?

Dr. Jeffrey Long: *First of all, I have a lot more confidence in your listeners. I think they have a lot more of a scholarly open-mindedness than [I heard in] your discussion with Dr. Woerlee. I have confidence in the substantial majority of people. That when they hear evidence and it's presented very straightforwardly, they're smart enough to understand what's real evidence and... what's basically evasiveness...*

...I admit, it's a little bit frustrating when people don't directly respond to the evidence. It's very carefully documented, referenced, and even corroborated with scores of prior studies. I think the best we can do, and you've been doing this for a long time, Alex, is just simply, calmly bring forward the facts, discuss exactly what the issues are, respond to any reasonable arguments that anybody brings up against the lines of evidence, and just go forward.

Me: *Let's touch on statistics for a minute because this is another rat's nest that I encounter with folks who have one position or another. Sometimes people want to grab onto statistics, but other times, like in this case, they seem to want to run from them. In your normal medical research, at what point do you reach a level where you say, "Gee, it's overwhelmingly suggestive that this is what's happening." How do you ferret all of that out?*

Dr. Jeffrey Long: *I think a lot depends on the number of patients studied and how strong the statistical evidence is. At other times it's just simply obvious by inspection. For example, when we... compared NDEs under general anesthesia and all types of causes of near-death experience, in 32 out of 33 elements studied there was no statistical difference between the two groups.*

Now, virtually anybody in the science or medical field would say, "Well, that pretty much nails it down that these two experiences are basically the same..." So it just becomes a matter of a judgment call.

But in fairly extreme situations where we have such strong evidence like [I've] presented in the book, Evidence of the Afterlife, *I* think that you have to rely on the good judgment and intelligence of your listeners and the world as a whole. Certainly from the feedback I've gotten from the book, the good news is that virtually everybody gets the statistics in the book as exactly the evidence that they are.

Later, in discussing some of the other lines of evidence for survival of consciousness after clinical death suggested by his research of near-death experiences, I asked him whether NDEs could occur in patients with little or no blood flow to the brain:

Me: *...It's medically inexplicable to have a highly organized, lucid experience while unconscious or clinically dead. Dr. Woerlee's [counter-argument] was that blood flow to the brain, mainly caused by CPR... explains why people are having these lucid experiences.*

Dr. Jeffrey Long: *There's a huge difference between simply blood flow to the brain and consciousness. I think that's where Dr. Woerlee's argument falls apart. Clinically, it's known medically that there's no substitute for talking to patients. When you talk to the patients who have actually survived CPR, one thing that is very, very obvious is that the substantial majority of them are confused or amnesic, even when they're successfully recovered. They may be amnesic for the period of time following their successful resuscitation or even for events prior to the time of their cardiac arrest.*

Me: *Can I interject something there? I want to stop you there on a minor point that you just breezed over. It's really important, and that's back to statistics... so statistically this group is supposed to [experience confusion and amnesia]... and they don't, right?*

Dr. Jeffrey Long: *Absolutely. If you read even a few near-death experiences, you immediately realize that essentially none of them talk about episodes of confusion or altered mental status when they just don't understand what's going on. You really don't see that at all.*

Again, for near-death experiences, they're highly lucid, organized events. In fact, in the survey we did, we found 76% of people having a near-death experience said their level of consciousness and alertness during the NDE was actually greater than their earthly, everyday life. So again, getting back to statistics, that's 3/4 and a substantial majority of the remaining 24% still had at least a level of consciousness and alertness equal to their earthly, everyday life.

So for that to be the statistics that you consistently see during near-death experiences and balance that with a substantial majority of people being confused around the time of their successful resuscitation from CPR, you really have to come away with the conclusion that even if there's blood flow to the brain induced by CPR, it's... by no means correlated with clear consciousness and alertness with near-death experiences. You just don't see that.

But also, in addition to that, note that the substantial majority of people that have a near-death experience, and have an out-of-body experience associated with cardiac arrest, are actually seeing their physical body well prior to the time that CPR is initiated. Once CPR is initiated, you don't see any alteration in the [narrative] of the near-death experience...

Me: That's a great point that I want to draw a little bit further because if we were to break down this process that someone's going through, there are three stages, right?

One, the brain is going dead. [For example,] they've had a cardiac arrest which [is the medical condition] we always talk about because we understand the physiology better than in other cases. [In the case of cardiac arrest] we know that within 10 to 15 seconds after their heart stops, their brain stops.

[Second,] there's some period that we have to assume that the brain is dead. [Finally,] there's some period where the brain is coming back online.

And what you just said is really important, and gets glossed over... [the information reported to you by NDE survivors]... seems to be consistent during the dead time and the coming back online time, right?

Dr. Jeffrey Long: Absolutely. There's no doubt about that. When

Wrong About Near Death Experience...

there's a cardiac arrest, the out-of-body observations that are often described during these near-death experiences certainly correlates to a time prior to CPR being initiated, and prior to a time there should be no possibility of a conscious, lucid, organized experience. And yet that's exactly what happens.

I'll tell you another thing, too, [if they were recalling their CPR], I tell you that you would hear a lot more from near-death experiencers; they would talk about their remembrance of the pain of the chest compressions.

Alex, that's a fairly painful procedure. It often breaks ribs and hurts. And yet, even when you have a patient who had a cardiac arrest and had a near-death experience, essentially never do you hear them describing as part of their near-death experience the pain of chest compressions. Again, the evidence just simply does not support Dr. Woerlee's speculations.

Me: And that rolls right into the second point of your nine lines of evidence in Evidence of the Afterlife... [near-death experiencers] see and hear in the out-of-body state—and what they perceive is nearly always real.

Dr. Woerlee's speculation was that maybe these folks are just seeing and hearing things with their normal senses through their normal bodies. One of the points you make is they're not experiencing pain. That seems to be more consistent with this idea that their consciousness truly is outside of their body.

Dr. Jeffrey Long: *And if their consciousness was really returning during CPR, wouldn't near-death experiencers not have out-of-body perceptions but describe their perceptions from within their physical body? And yet you don't see that with near-death experiences.*

So in other words, if you started CPR and they had a near-death experience and suddenly they started to have some consciousness, you'd expect that instead of having the out-of-body experience where their consciousness is apart from their body, their consciousness would be within their body. You just don't see that.

Me: ...What about the point he makes about the two types of out-of-body experiences? [Ones that are] relatively close to their body and the ones where there is this distant gathering of information down the hallway or miles away. Any thoughts on that?

Dr. Jeffrey Long: That's an interesting point. If you want to arbitrarily divide out-of-body experiences, that's one way you could do it. Most out-of-body observations during NDEs are of events going on around their physical body.

Yet there are dozens of these out-of-body perceptions during near-death experiences where they can hear and see events far, far removed from their physical body, often in completely different rooms, geographically far away, where any possible physical sensory awareness should be absolutely impossible.

And yet when they make these very remote out-of-body experience observations, their accuracy is absolutely the same—about 96% — as the observations of events going on around their physical body...

Me: You mentioned that that distinction would be arbitrary. You're saying from the data you've seen, you don't see any reason to separate out, based on the accounts, based on the information you're getting back, there's no substantial difference between people who are describing events close to them and people who are describing events further away from them.

Dr. Jeffrey Long: They're both highly accurate. I mean, how far of a geographic distance would you want to part from the body before you're going to call it a distant OOB observation? No one's really made that distinction in the NDE literature. So really, the bottom line is it's just another strong line of evidence for the reality of OOB [out-of-body] observations. They're equally accurate, even if these observations are far, far away from their physical bodies.

Me: Okay. [Dr. Woerlee]seems to be suggesting they're of a different quality, that they're not describing things in a different way or more hallucinatory. It's pretty much the same across the board.

Dr. Jeffrey Long: *In fact, it's often the case that they'll make out-of-body observations of events right around their physical body during the NDE, and then as part of the same experience, make out-of-body observations far removed from their physical body. Absolutely no difference in what they're describing.*

Me: *Let's talk about "blindsight." This is another interesting point. In your research you observed that NDEs take place among some folks who are blind, even folks who are blind from birth. Now Dr. Woerlee's counter-argument to that was 1) maybe these folks are getting visual information from the people around them who are describing things, but then 2) he also seems to suggest that many blind people are extremely good at creating these mental maps of their surroundings.*

I found that a little bit of a stretch. Do we really expect that folks who are visually impaired or completely blind from birth would be able to describe visual information as well as sighted people?

Dr. Jeffrey Long: *That was puzzling to me, too... all of a sudden, for people blind from birth, the first time they have vision is while they're unconscious or clinically dead? I mean, it really doesn't add up.*

But certainly the people who are blind may get some visual information from people around them, but when you talk to people who have been blind from birth [and then] have a visual near-death experience... this vision is a unique perception to them. No matter what they've done in the past as far as mental maps or gathering visual information, whatever that means, the actual ability to see during their near-death experience is a new experience to them. It's clearly different from anything else going on in their life.

Me: *Let's talk about a couple of other points that... I think are really fascinating. Number one is the family reunion: Your data suggests that people consistently meet loved ones in the afterlife, and that number two... 98% of the folks they encounter are deceased. I didn't really hear a good explanation from Dr. Woerlee on that.*

Dr. Jeffrey Long: Let me back up and define what we're talking about. When people have a near-death experience, they may have out-of-body observations and see ongoing earthly events—that's not what really counts in this [case]. You have to be in an unearthly realm and you have to be seeing people that you knew during your earthly life that are not going about their daily earthly events, like in an out-of-body experience observation. So these are purely observations in unearthly realms.

You're right. The percentage of time that people encounter deceased relatives is extremely high. It was actually 96% in the NDERF study and only 4% of near-death experiencers met beings who were alive at the time of the near-death experience. That's actually corroborated by another major scholarly study which found it was 95% of the time that they encountered beings they knew from their earthly life that were deceased.

The important thing is that any other experience of altered consciousness that we experience on earth, dreams, hallucinations, drug experiences, you name it; in all of these other types of experiences of altered consciousness a vastly higher percentage of people are going to be alive at the time of their experience.

You're going to remember the banker that you did business with that day, or your family member you said hi to as you were walking into the house. This is what's in the forefront of consciousness. So for people to so consistently encounter deceased relatives is very, very strong evidence that they are, indeed, in an unearthly realm and it certainly points to evidence of an afterlife.

Me: Any thought on the striking similarity of content among very young people who have experienced a near-death experience, or the remarkable consistency of near-death experiences around the world?

Dr. Long: A really interesting part of the study that I did was looking at children ages five and under. In fact, their average age was 3 ½ years old. These are children so young that to them death is an abstraction. They don't understand it; they can't conceptualize it. They've almost never heard about near-death experiences, have no preconceived notions about that. They certainly have had far less cultural influence, both in terms of religion or anything else that could even potentially modify the near-death experience at that tender young age.

And yet looking at these same 33 elements of near-death experience that I did in other parts of this study, I found absolutely no statistical difference in their percentage of occurrence in very young children as compared to older children and adults. That almost single-handedly shoots down the skeptical argument that near-death experiences are due to pre-existing beliefs or cultural influences. We're not seeing a shred of evidence that corroborates that at all. In fact, that finding is actually corroborated with another major scholarly researcher who actually reviewed over 30 years of near-death experience research and came up with the same conclusion.

Children, including very young children, appear to have identical near-death experiences as adults. That seems to be a fact. And that's not explainable by any medical, psychological, or expectation theory that could be advanced.

Me: Let's spend a minute talking about the last point in your nine lines of evidence, and that's "Changed Lives": NDErs are transformed in many ways by their experience, often for life. I think Dr. Woerlee suggested that a brush with death [on its own] could change anyone's life...

Dr. Long: There have actually been two good prospective studies in which there were cardiac arrest survivors. Most, of course, did not have a near-death experience, but a number did. And these researchers followed the two groups longitudinally over a period of many, many years. In both studies, they found that there may be some life changes in people that had a heart attack, but there were far more life changes among near-death experiencers.

So it certainly looks like it's the near-death experience, and not *just the life-threatening event, that accounts for the preponderance of the changes seen after a near-death experience. I think that's pretty well established.*

All other hallucinatory events, dreams, all other temporary, transient, even pathological alterations of consciousness are essentially never going to result in such a high a percentage of people experiencing those types of profound life changes that we see in near-death experiencers.

And moreover, what you see in the life changes of near-death experiencers is markedly consistent. In other words, it's not just that they have life changes; it's the consistency of those life changes. The substantial majority, if not overwhelming majority, of near-death experiencers believe that there's an afterlife. They believe that there's a God. They no longer fear death. They're less materialistic. They value loving relationships more. The list goes on and on. This has been consistently observed not only in our study but in scores of prior scholarly studies of this phenomenon over 30 years.

I wrapped-up the interview by asking Dr. Long whether the kind of evidence he had spent so much time collecting could ever change the mind of hardcore, science-as-we-know-it types:

Me: *...It was the data that convinced you. You didn't have a NDE. No one close to you had a NDE. Do you want to speak to that process of change that you've seen people go through?*

Dr. Long: *As a physician that fights cancer, I'm very evidence-based. In fact, I require strong evidence before I'm going to make any therapeutic decisions in my professional life. Therefore, regarding my understanding of near-death experience, I had exactly the same standards. I'm a real show-it-to-me kind of person, anyway. So I really had to see evidence, and I had to see overwhelmingly strong evidence, before I was going to be convinced. The evidence is compelling for the reality of near-death experience and it's consistent message of an afterlife.*

The end of my conversation with Dr. Long had brought me

back to the question I kept encountering throughout all of the different topics I investigated on *Skeptiko*: How could this be? How could otherwise intelligent, competent, seemingly honest people be locked into a mindset that kept them from the kind of open-minded, objective, rational thinking they advocated? In the next chapter we look at how this defend-the-status-quo thinking has become deeply ingrained in the scientific establishment.

CHAPTER 5: WRONG ABOUT PSYCHICS AND MEDIUMS (Part 1)

[Scientists and specialists] are ultimately human. They may have vested interests. They certainly have egos. They may be concerned about saving face, or obtaining funding or tenure. Or they may simply have prejudices, areas of ignorance, or experiences which bias their thinking. While respecting them for their superior knowledge and experience in an area, we should never impute superior judgment to specialists.[20] — Astronomer Tom Van Flandern

I remember the first time someone called me an iconoclast. I quickly ran to Google to look up the meaning. Sometimes you don't know what you are until someone tells you. My buck-the-trend sensibility has drawn me to researchers willing to do the same. When it comes to doing serious science of a widely ridiculed phenomenon like psychic communication, you need a thick skin. You also need a determination to tune out a mainstream scientific community well known for dismissing a phenomenon that offers a direct challenge to the mind equals brain definition of consciousness.

Although I had never had a psychic reading or visited a medium before starting *Skeptiko*, I knew from personal experience that most people were open to the reality of this kind of anomalous commu-

nication. In fact, I've yet to find someone who doesn't have a story of "visitation from the beyond" somewhere in their family history. But I was still struggling with the "how can this be?" question. I mean, if these paradigm-busting experiences are so common, why aren't PhD-types lined up ready to take on the challenge and pick up an easy Nobel Prize? And even more surprisingly, why do so many mainstream scientists take a firm, even hostile, position against the phenomenon? I've discovered three reasons why this is so:
1. They are willfully ignorant of the research that exists;
2. They never personally investigated the topic themselves; and
3. They can't accept any anomalies that challenge their carefully constructed mind equals brain paradigm.

Dr. Julie Beischel

My *Skeptiko* exploration into psychics and mediums began in 2006 when I contacted Dr. Gary Schwartz, a professor of psychology and medicine at the University of Arizona. Dr. Schwartz caused quite a stir when he started carrying out experiments with mediums in the early 2000s. I had heard about his research and wanted to learn more.

Schwartz presented stellar credentials: a PhD in psychology from Harvard University, a faculty member at Harvard after only five years, and past chair of the department of psychology and psychiatry at Yale. But as he started to publish his findings on medium communication in professional journals and popular books,[21] he quickly became a target for those with a knee-jerk reaction to this type of research. Predictably, they started their attacks with accusations of "cold readings" and suggestions that this career researcher, who had published over 400 scientific papers, had stumbled over basic, easily-controlled-for research problems involving the blinding of subjects and controlling for information leakage.

The debunkers spared no effort in assailing Schwartz. This culminated in a sensationalized 2007 Geraldo Rivera television report

Wrong About Psychics and Mediums (Part 1)

calling into question his fundraising. It was then that Schwartz decided to end his research into mediums and move on to other areas of research.

Fortunately for me, this is where my story begins. Through my interaction with him, I was introduced to his associate, Dr. Julie Beischel, who, as it turned out, was the person who did much of the heavy lifting when it came to the University of Arizona's medium research. Her Ph.D. in pharmacology and medical research expertise made her extremely well qualified for setting up testing protocols. Schwartz acknowledged relying heavily on her. After Schwartz stopped doing medium research, Beischel established the Windbridge Institute, a research organization dedicated to serious scientific investigation of medium communication for the grieving.

The following excerpt comes from an interview I had with Beischel in August 2008 (*Skeptiko* #51). This was early on in my *Skeptiko* educational process, when I naively thought that all it took to convince the mainstream scientific community was good, solid research. Here, Beischel and I discuss her research protocols and the measures she's taken to address skeptical claims that medium readings can be explained through the use of stage-magic-style "cold reading" techniques.

Dr. Julie Beischel: *If the medium and the sitter are in the same room, obviously there's the potential for sensory leakage. For example, even if you use a partition, the tone of voice can give a lot of information to the medium. So you can't have the medium and the sitter in the same room. Even on the telephone there's sensory leakage. So in the current research the sitter is not on the phone with the medium. And as the "proxy sitter," I'm blinded from any information about the person who requested contact with the deceased (i.e. the sitter), or the deceased person.*

Me: That's great. And I want to make that crystal clear. On the readings that you are doing now, the medium never talks to the person they're doing the reading for. Is that correct?

Dr. Beischel: *That's correct.*

Me: *So that part of the argument about the use of cold reading techniques is totally out of the equation. And I want to point out that the counter-claim to that has never been proven. I mean, it's never been proven that non-verbal cues and tone of voice can explain all the information given in a reading.*

Dr. Beischel: *I would agree with that.*

Beischel and I then examined other types of problems associated with trying to design well-controlled medium experiments. We also discussed the criticisms of skeptic Ben Radford, a science writer who had appeared on *Skeptiko* to make all the usual arguments as to why this kind of research should be ignored:[22]

Me: *Let's move on and very quickly go over your protocol, and then it will become really clear to anyone who's objective that there cannot be any judging bias and there cannot be any control group bias.*

Dr. Beischel: *Our current protocol uses a quintuple-blind methodology, so there are five levels of blinding.[23] How we do it is we start with a group of sitters and we screen them, and they describe the one person they wish to hear from most that's deceased. Then we take those descriptions and we find pairs of deceased people who are the most opposite, and that creates a pair of sitters. Then each medium reads a pair of sitters and a pair of hypothesized "discarnates."*

Me: *One of the things that's different from the prior research that was done at the University of Arizona is you narrow it down and say, "Okay, I want you just to select one person that you specifically want to talk to on the other side, if you will, one deceased person."*

Dr. Beischel: *That's correct.*

Me: *And you actually pre-select pairs of people based on how different the person that they're trying to connect with is. So take us through an example of how that would work.*

Dr. Beischel: *We pair the people to be most opposite in age,* *physical description, personality description, hobbies, and cause of death.*

Me: Okay. So I come along and I'm a potential research participant, and I say, "Gee, Julie, I'd really like to connect with my grandmother who passed away. She passed away when she was 95, and she was a small woman in stature, five foot one, and she had olive skin, and her hobbies were going to church and knitting."

Dr. Beischel: *We would find another sitter who had a discarnate who was young and active and blonde and had died of an accident as opposed to your grandmother who died of old age. So we pair the person to be most different on a variety of characteristics: it's hair coloring and build and height, etc.—and cause of death would be, for example, the part of the body that was affected, fast or slow, natural or unnatural...*

Then the same medium reads both people in the pair. If you said you wanted a reading from your grandmother and I wanted a reading from my sister, then a medium would do a reading for your grandmother and one for my sister. Then I would score both readings and you would score both readings, but we wouldn't know which one was which, because we weren't there when the readings took place. This controls for reader bias. Next, we give the readings a score. I give a score to each reading and you give a score to each reading, in addition to item-by-item scoring. It's a very complicated scoring process. So my score of your reading serves as a control, and your score of my reading serves as a control. Then we statistically compare the scores given by the intended sitter to the ... scores given by the control sitter...

Me: *Awesome. Now let's back up because, actually—and I know you're trying to make it as simple as possible—there's a couple of other steps that you go through there that are also interesting and noteworthy. So we'll put you in the place of the person who's talking to the medium. The first thing you say to the medium is: you give the first name of the person you're trying to connect with, right?*

Dr. Beischel: Yes. So, during the reading, all the medium and I have is the first name of the discarnate. So we start the reading, and I give the first name.

Me: Great, and that's a good point, too. That's all you have, so you don't have any information because we're substituting "you" here as a pronoun, but it's not really you. You wouldn't be the person who decided who the two participants were or who the two deceased people that we're trying to connect with. You would be blind to that.

Dr. Beischel: Right. A different experimenter does the pairing and screens the sitters. It's my research assistant, Michael. So Michael gets on the phone with all the sitters. He gathers all the information; he does all the pairing. And then I say, "Okay, I'm ready to do two readings," and he gives me a first name.

Me: So you sit down there and you say, "Medium, I don't know anything else other than I have a person here who wants to connect with someone named Sarah. Ready, go."

Dr. Beischel: Let's back up even a little more. During the reading, I give the medium the first name and they're allowed to just give some general information for about ten minutes, and then I ask those same four specific questions: describe the physical appearance of the discarnate, describe their personalities, what were their hobbies or how did they spend their time, and what was their cause of death?

Me: Now the reading ends. What happens next?

Dr. Beischel: Then I'm still blinded. I still don't know anything, so I take the recording of the reading and I turn it into a list of single individual items of the information that the medium provided.

Me: Great. But just to be clear: if I said you're doing a reading for Sarah, you would take out obviously any references to the name Sarah because that would tip off that that's who the reading is for.

Dr. Beischel: That's exactly right.

Me: So now let's get down to the payoff. Now you've done a couple of readings with you being the intermediary. You've transcribed them, broken them down into a list of declarative statements: "Her hair is red. She passed away this way. She was introverted." Now what happens?

Dr. Beischel: So now I send those two readings to a third experimenter, Mark.

Me: So Mark gets the two readings, and he sits down with me and says, "Okay, Alex, you wanted to connect with your grandmother, Sarah. Here are two readings. I don't know which one is for your grandmother Sarah, but one of them is and one is for another person." Then I'm asked to do what?

Dr. Beischel: He doesn't sit down with you. He emails the readings to you, and Michael has previously trained you on how to do the scoring. So Mark sends you two readings. You already know how to do the scoring, and by yourself you score each of the readings, and you email back your scores.

Me: Great. And what about the mediums who are giving a lot of generalities to their answers and are just fishing around for information? How does that not affect this process?

Dr. Beischel: Well, they can't just give general information because we're asking for specific pieces of information. If they just say, you know, "He's kind of short but kind of tall," that's two items now: he's kind of short, and he's kind of tall. So a medium that just provided general information would have a low accuracy percentage, and the sitter would have a lot of trouble discerning between the two readings, so they probably would have a low record of the sitter's choice.

Me: So in your published, peer-reviewed study that follows this protocol, what percentage of time did sitters choose the correct reading?

Dr. Beischel: 81 percent. So 13 of those 16 people.[24]

Me: And that's pretty impressive. That was statistically significant if I remember correctly from the paper.

Dr. Beischel: Yes, it's quite significant. You're looking for a P value of less than .05, and I think that P was .001.

Me: *I think the whole process is extremely tight. It sure seems to me like it's addressed the main issues. With that in mind, let me go back to my interview with skeptic Ben Radford:*

I then played this audio clip for Dr. Beischel.

Ben Radford: *I do know about the research, and one of the issues it doesn't address is that in many of these cases the verification of the information is provided by the sitter. That is, this is not information that's supposedly coming from the great beyond—that is verified by a third party person. Much of this is information where the medium will say, "I'm getting information from your husband or grandfather or whoever else," and the information is judged either accurate or inaccurate by the sitter, and there is an inherent problem right there that has not been addressed.*

And, I added one more question after I played the clip:

Me: *I just want to remind people that I specifically referenced your research, and he responded, "Yeah, I'm familiar with that research." So what are your thoughts on that?*

Dr. Beischel: *He's listing as a criticism the fact that the sitter is the person that judges the information as accurate or inaccurate. So it's important to keep in mind the scope and the goal of the research. The goal isn't to prove the existence of an afterlife. What we're doing is examining the processes of mediumship in its natural environment, with the proper controls—in other words, normal readings between a medium and a sitter.*

Again, the general hypothesis we're testing is: can mediums report accurate and specific information without any prior knowledge and in the absence of any sensory feedback? So with that being the goal, the sitter has to be the person that is the judge because the information was intended for the sitter. We don't ask the hypothesized discarnate to take an algebra exam and provide information a third party could determine is accurate. That's not what a mediumship reading is. We're asking them to communicate with their friends and family, the only people who can determine if the information is correct and applicable.

Me: *Okay, that's a very good point.* Just to clarify: where I think Ben was coming from is that if a person comes through and says, "I think I'm connecting with an aunt figure and her name is Dolly," then the person who is accepting the reading can say, "Yeah, yeah, that's right." Then there isn't the independent verification of whether that's right or wrong that was troubling to some skeptics.

Dr. Beischel: *We're controlling for reader bias by having the person score two blinded readings, and we're comparing their score of a reading intended for them to their score of a reading not intended for them.*

Me: *So the big problem the skeptics point out, and in some cases very correctly, is that the reading could pile up like positive points, like, "That's a hit. That's another hit." Now what you've done is kind of taken that out of the equation, because it really doesn't matter how high or how accurate any particular reading is, it's more of a comparison: how does this reading compare with the other reading? They can be biased one way or another way, but it's going to wash out with the control reading that they have.*

Dr. Beischel: Right. There will be the same bias for each reading because they won't know which is theirs.

Me: *Here's the next quote from Ben:*

Radford: Part of the problem here is that those descriptions that you just gave can still be vague. Someone says, "You know the person who's coming through is a tall man with gray hair." Well, it turns out that when the person died, he was bald. This gets back to the problem of having a sitter verify the information. Because the sitter says, "Yes, he had gray hair." Then that's a hit, that's good information. But the medium could also say he was bald, in which case the sitter would say, "Well, he had gray hair, but in his last years he was bald." So you can have a medium giving two contradictory pieces of information, both of which would be considered a *hit by the sitter.*

Dr. Beischel: *I think this other point is important to address and then I'll get to Ben's argument specifically. Ideally, laboratory-based mediumship research has to include two things: 1) an environment that optimizes the process for everyone involved—the medium, the hypothesized discarnate, the sitter—in order to increase the probability of capturing the phenomenon if it exists, and 2) methods that maximize the blinding to control for any conventional explanations for the information. Together, those two factors optimize the possibility of achieving positive results while also controlling for experimental artifacts.*

I have a real world example that will make that much more understandable. We use this metaphor: you can't study football on a basketball court using baseball players and the rules for hockey. Because if you get negative results, you can't say, "I've disproven the phenomenon of football"—because you're on a basketball court using baseball players and the rules for hockey. That's not a proper experiment. Similarly, it's not appropriate to claim that Jason Elam could kick a 95-yard field goal if you give him a Nerf football, an empty stadium and no defensive line. That's not real football either. In order to study football appropriately, only trained skilled participants and the regulation equipment, environment and regulations can be used. The same thing is true for mediumship...

Me: *Right. Once you put the proper controls in place, why not do everything possible to recreate the phenomena? I completely agree. Now what about Ben's point about the information being vague? Maybe [the medium] says, "The hair was black, but [also says he] was bald."*

Dr. Beischel: *I think his main point was that two pieces of contradictory information can be judged as hits by the sitter. Again, we have to keep in mind the process of mediumship. So first, the medium is interpreting images, symbols, sounds, whatever, so she actually may receive two items that appear contradictory, but we instruct*

Wrong About Psychics and Mediums (Part 1)

them to say what they see. That's the process, and we're studying the process, so we can't mess with it. Second, as is often the case, as anyone who has ever been in a relationship can attest, communication between people is not black and white. Things can be contradictory and accurate at the same time.

Me: So then let's follow those seemingly contradictory pieces of information through your protocol and see if they do create this mix up in the final judging of the data. So the person says, "Black hair, bald," two things that [appear] contradictory [but may both be true if] older in life they were bald, and earlier in life they had black hair. How do you score it? How do you report it? And then how does the sitter score it?

Dr. Beischel: We're not asking the medium to describe a photo of the discarnate. We're asking them to describe aspects of the entire dynamic lifespan of the person. It's entirely accurate for the person to be bald at some points in their life and have dark hair at other points of their life—and the discarnate may present themselves to the medium in each of those ways. It's also accurate for the sitter to score each of those items as accurate because they are both accurate. And during scoring, again, we're not racking up the number of hits. We're comparing the number of hits to the number of hits in a control reading.

Me: Which is a really important distinction. Okay, here's the next clip from Ben Radford's critique of your work:

Radford: I'll tell you what: use subjects who do not fit what most people would consider to be a normal profile. Have a subject who maybe lost his legs in an accident during the war.

Me (to Dr. Beischel): I think we've handled this already.

Dr. Beischel: I did actually want to address that specifically. Again, we have to keep in mind the scope of the research. We're not setting up a condition where the discarnate has to prove something. We're interested in studying mediumship under conditions that exist normally. So that is where the medium provides information allow-ing the sitter to identify the discarnate.

Me: *I think where he was going was just to say, "Let's get people that are different so that those differences are highlighted in the reading," and that's the first thing that you do! I just want to bring us back to that and make it clear.*

But let me go ahead and really go out on a limb here and speak for Ben Radford. I think where he was going was just taking the mundane out of it.

Dr. Beischel: *I have one word: football. That's like saying, "Oh, I don't like watching football because the players don't fly." Well, that's not how football works, so we have to keep in mind how the process works. We've actually found that mediums most often report three kinds of information: 1) information that allows the sitter to identify the discarnate—so I call that, "It's me! It's me!" 2) events that have occurred since the discarnate's passing—"I'm here! I'm here! I'm still here. I saw you at that birthday party" or "I saw that you got married. I'm still in your life even though I've died"; and 3) messages of an emotional nature, like "I love you." Again, we're not looking to prove the existence of an afterlife. We're interested in the phenomenon of mediumship. Mundane or not, the information is meaningful to the sitter, and that's what we're studying.*

I think it's also important to recognize that the majority of all human communication isn't mundane. I like to say we study human communication—one of the people just happens to be dead. You don't study normal human communication by asking people to talk about quantum physics. You just ask them to talk to each other, and that's what we're doing.

Me: *And, of course, what's mundane for one person or another, or a third observer, is incredibly relevant to an individual. I can say from my own personal experience, some information I was given would seem incredibly mundane, but it was deeply, deeply meaningful to me.*

Dr. Beischel: *I think that brings up another point. Mr. Rad-*

Wrong About Psychics and Mediums (Part 1)

ford said something about the discarnate saying, "I love you," and he asks, "What value is that?" Again, let's remember what we're studying. It's a deceased person communicating with the people that they loved, who are grieving and suffering, [this is] not a graduate student defending their dissertation. So what would you say to your family if you had died and you could witness their mourning? I would say, "I love you."

I began this chapter with three reasons science hasn't tackled medium communication. The first is ignorance of the existing research. This interview with Dr. Julie Beischel shows the kind of solid, scientific research naysayers continue to ignore, distort, and misrepresent. In the next chapter, we'll look at why scientists remain stuck—they never personally investigate the topic.

CHAPTER 6: WRONG ABOUT PSYCHICS AND MEDIUMS (Part 2)

Something has gone wrong...we can't read outside and can't hear inside. So I say we have become idiots...the origin of the word idiot is the word private. Each of us has become private and no longer shares the thought of the [world] brain...thus our real life and purpose is conducted below our threshold of consciousness.[25]

— Author Philip K. Dick

My Dad was my rock, my role model, my hero, but when he died in 2000 I didn't do a lot of grieving. My Dad suffered from Alzheimer's for years prior to his passing, and as anyone who's experienced this disease knows, you grieve for your loved one long before they pass away. When I began to investigate medium communication on *Skeptiko*, I knew I would eventually try it for myself. And there was one person I wanted to connect with—my Dad.

I took a straightforward approach when investigating medium communication and trying to connect with my Dad. I began by scheduling a reading with one of the mediums Julie Beischel recommended from her research. Next, since I was aware of the kind of basic controls one needs to have in place to avoid "cold reading" problems, I instructed my medium that I would be providing very little feedback during our session—only simple "yes," "no," or "I don't know" answers. She agreed.

The reading was a bust. Nothing about the information she brought through seemed to fit my Dad or anyone I knew. I was disappointed, especially since the medium told me that this lack of connection was very unusual. Then, contrary to everything I had heard about medium scams, she quickly offered a full refund.

I tried again. My second reading was with a highly regarded medium I discovered through an internet search. He was a likable guy, but the results were similar. And again, without prompting, he insisted on giving me a full refund because of our lack of connection; so much for psychic rip-off artists. At this point I was wondering if there was something wrong with me. I had interviewed researchers and experts claiming great success connecting with the "other side," so why wasn't it working for me?

I decided to try one more time and this time had an amazing phone reading with Willa White from the spiritualist capital of the United States: Lily Dale, New York. The connection with my deceased father was undeniable. White not only described his physical characteristics and personality, but relayed some personal information no one else could have possibly known. I was excited and a little surprised to have finally experienced it firsthand. I had personal verification of the reality of this phenomenon. I wanted to find a way to share my experience with *Skeptiko* listeners.

Marilynn Hughes

I decided to arrange a public demonstration of properly controlled medium research on *Skeptiko*. For purposes of the show, I had decided to square off against a group of high profile skeptics who insist all psychic communication is fraudulent. My would-be associate was Dr. Steven Novella, a neurologist and assistant professor at Yale University. Novella is very popular within the skeptical community and is the host of *The Skeptics' Guide to the Universe* podcast. While Novella and his colleagues showed enthusiasm towards the demonstration during our initial interview and planning stages, they lost interest as the ultimate showdown neared (*Skeptiko* #44). It's a pattern I've seen repeated many times since.

However, after months of research and over a dozen "proxy readings" with mediums, I was too far in to abandon the project. Something else drove me forward as well. During my research into medium communication I had not only become convinced of the reality of the phenomenon, but I had been introduced to a deeper purpose of medium work: helping to relieve pain and suffering of those in grief. With this goal in mind, in June 2009 I published a medium demonstration with a grieving mother in Austin, Texas (*Skeptiko* #75).

The medium I chose for the demonstration was Marilynn Hughes. I had interviewed Hughes while recruiting mediums for the demonstration I was planning with Steve Novella and the skeptics. I liked Marilynn, but her style was different from the other mediums I had spoken with. In fact, she didn't even consider herself to be "medium." Hughes claimed to use "out of body travel" to contact deceased people. It was a "fringy" claim, even for those who study mediums.

Marilynn Hughes: *When we leave our bodies and we go into different spiritual states, we enter worlds that have different laws than third-dimensional reality. So in order to take science into that study, somehow we have to integrate and include those different laws of existence.*

"Different laws of existence?" Come on! Isn't this just the kind of stuff that keeps serious researchers from investigating this topic in the first place? I started to worry about maintaining respectability. On the other hand, what if there was some reality to it? The only way to find out would be to do a reading. I asked Hughes how we might go about setting up our test.

Hughes: You know, one *thing I think is workable would be utilizing out-of-body experiences for messaging people. And I don't know how far we can take it.*

Me: *That's awesome. So let me make sure I understand; you're saying spirits come to you and say, "Marilynn, could you please help me with connecting with so-and-so?" And then, in your experience, a lot of times those messages are answered, and then you pass those along.*

Hughes: *Right. And that's why I'm saying I'm not sure how far we can go with it in a situation like that, because it is God-directed rather than me-directed.*

I still had reservations about the "God-directed" part, but I wanted to follow Dr. Julie Beischel's advice and allow my medium to work in the way they felt most natural and comfortable. We eventually agreed on a set of simple protocols. I would select a person who wished to connect with a deceased loved one. Hughes would not be informed of the participant's identity, and this person would only communicate with me during the readings. I would be acting as the "proxy sitter"—to use Dr. Julie Beischel's term. I would send very limited information about the deceased loved one to Hughes, after which she would email me the readings. Finally, I would send the reading to the person who had requested it.

Megan

Because of my previous research, I had built a database of potential participants (i.e. "sitters") to choose from. I contacted a woman named Michelle from Austin, Texas, and she agreed to participate in a public demonstration. Tragically, Michelle had lost a daughter to suicide at age 20 and very much wanted to reconnect with her. Following the protocol Hughes and I had set up, I asked and received an item of clothing from Michelle's daughter and some photographs without any people in them. I also collected her daughter's first

Wrong About Psychics and Mediums (Part 2)

name (Megan), the day and month of her birth (but not the year), and the date of her passing.

I emailed Marilynn and told her she was going to try and connect with someone named Megan. I also gave her the month and day of Megan's birth, but not the year. As I was getting ready to send her Megan's hat, but before I even got it in the mail, Marilynn astounded me with her first reading:

Megan is a soul who definitely wishes to make contact with her family, and apparently has some level of permission from God to do so. I have some random things to share that she showed me. I don't know what they mean. Let the family see if they mean anything.

I should make it clear here that, throughout the entire time that Hughes relayed her readings to me, I never told her whether anything she was telling me was accurate or not. Here are some of the statements she included in that first email:

First, she showed me what appeared to be a college environment.

This was correct. Megan had been in college right before she had committed suicide. The next section was definitely not something I was expecting:

I don't know what this means, but I watched her begin on campus and then somehow get lost. She ended up not being on campus but I didn't feel like she was that far away. She wandered off with a woman who was definitely, very clearly lesbian. It seems that Megan wasn't sure for a very, very short time of her identity, but she soon realized that she was not a lesbian. My sense was that she was definitely showing me a time in the late teens, early twenties.

As you'll soon see these facts were stunningly, unexplainably accurate. After receiving Megan's hat, Marilynn provided a second reading:

Megan told me that her death was very, very hard on her mom, and that she really loved her mom and had a close relationship with her.

This information was accurate but not astounding. Then again, not all mother-child relationships are close. And I never told Hughes it was Megan's mother who was trying to connect with her. It could have been her brother, sister, or a friend.

The next few bits of information were more specific and turned out to be very meaningful to Megan's mom. Hughes reported that Megan had been experiencing a returning home to her faith:

It was like she had been somewhere else for a while, maybe off to college or a different location. Maybe she just went away from the church, and she came back. And she was very happy to be back.

Then she went on to write:

It was kind of a calm before the storm. Yes, I am hearing from her a <u>calm before the storm</u>. She had thought she had found her way back home, but something else was going to happen. Then she said, "The woman told me a different story. Something she told me wasn't true."

This passage might sound a little cryptic, but it turned out to be very relevant. Megan's mom later told me that this was exactly the kind of spiritual crisis Megan had gone through. Prior to her passing, she had drifted away from her religious tradition, then she had found her way back. Her mom reported that she had joined a Christian Bible study group. The storm, of course, was her eventual suicide.

Meanwhile, Hughes mentioned that this case reminded her of a previous case where a family was trying to determine the cause of death. In this case, the uncertainty about the cause of death had created a lot of pain for the family. Hughes sensed there was something similar going on in the case of Megan's passing.

This turned out to be very significant. I later discovered it was the primary reason Megan's mother had contacted me. Michelle explained, "Megan's death caused a major divide in our family. Her death was clearly a suicide, but there is a part of my family that isn't willing to accept that she really killed herself. They think there are other reasons behind it. She was on anti-depressant medication; she had gone through a car accident and was on pain medication. And they want to believe that there are other reasons behind it."

Near the end of the reading an important nugget of information came through. Even though it was a small detail that would seem insignificant to an outsider, it was powerfully meaningful to Megan's family. It's illustrative of how readings often work in strange ways. Hughes wrote:

I see Megan, and she is smiling. There are these flowers all around her face. It is a strange symbol. I think these flowers are peonies.

At the time of the reading, Megan's mom had been talking to Megan's father about planting a garden to honor Megan. She asked him what kind of flowers she should plant.

After the experiment was over, I carried out a follow-up interview with Marilynn to discuss the process she had gone through.

Me: *So what was your process in terms of connecting with Megan?*

Hughes: *With Megan, it was almost instantaneous. I had asked for the date of death because sometimes it's very important and helpful when you know how long someone has been crossed over, because there are different stages in death. But with Megan, as soon as I got her name and you gave me those dates I could feel her. She was just all over me. And it was like, wow, this girl wants to talk to her family.*

The process is really not something that I control, but it's a process where, when I go to bed at night, I have these out-of-body experiences. And then the person who is deceased will show me things. And a lot of times they show you their moment of death. I believe this is what I experienced with Megan, based on what you just shared with me afterwards.

In the last reading, I think that's when she probably died. She was in a room, and I woke up bone-cold, and that's what it felt like:

oh, she must have died then. So the process is: you go out-of-body and the person takes you on a visual where you almost experience it with them. It's almost like you become them in the experience, where you are experiencing part of their life as them.

So you can feel their emotions, their feelings, everything that's going through their mind. And that's part of the process that is very important, usually for the remaining family members. Because a lot of times in a situation like this, the families blame themselves, and it's always helpful for them to know, if it's possible, what it was that led to the moment of their death. There are so many things involved in the process, but it's primarily something that is led—I believe it is led by God.

Me: So that's interesting—and it's also interesting that you're talking about God. And this is going to be very controversial and very challenging for a lot of folks, but these are just the facts of this case. One of the things that you reported in your reading was this struggle that Megan was having and how she had gone and drifted away from her spiritual path and had run into some pretty bad people along the way, and then she had found her way back on her spiritual path.

Hughes: She was back in a church and was back in a group, a religious group of some kind. The way she presented it to me it was the faith of her youth.

Me: Well, this is exactly what happened in Megan's case. While she was at college, she had befriended this woman who was really kind of a "bad news" person. This woman was somewhat of a psychic, but was running a psychic scam. This woman also had a Wicca background, whether that's good or bad.

Hughes: That explains a lot, yes.

Me: And this relationship was quite troubling to Megan and to her family, but her mother reports that near the end she had come back to her Christian faith and was trying to get back into that and was reading the Bible regularly. So again, these are just the facts,

Wrong About Psychics and Mediums (Part 2)

and they match perfectly with the reading that you gave, and there really isn't a good or reasonable explanation for why you would connect any of that up or why you would provide any of that information.

Hughes and I then examined the message she had received regarding the peonies.

Me: *I will tell you one specific thing that you gave that had tremendous meaning for Megan's mom, and that was the flowers.*

Hughes: *Really? She made a big deal about the flowers, and I was almost hesitant to even write it because I was like, "How can this even mean something?" I thought, okay, I'll just say it. So that's interesting. What meaning did that have?*

Me: *Megan's mom was in the process of building a flowerbed specifically for Megan, to kind of celebrate Megan. She had been talking to her husband about, "You know, I really want to start a little flower garden."*

Hughes: *Wow, I'm getting slammed! Now I understand. Because she had the flowers around her head and it didn't quite make sense. But now it makes perfect sense that the flowers would be surrounding her.*

Me: *And those specific flowers that you said.*

Hughes: *Peonies. Were they peonies?*

Me: *Megan's mom was so moved by this that she went and researched these flowers, and found that they were exactly the kind that she was looking for. They were indigenous to Texas, and they met all the criteria that she had laid out in her mind in terms of how to construct this flowerbed to honor her daughter.*

Hughes: *Wow, that's so exciting to hear that. One of the interesting things about doing an experiment like this is that most of the time when this sort of thing is happening I know a little bit about the family. So this was actually very interesting for me to do because of the fact that there was the "no feedback" rule. So it was kind of running blind and just hoping that you are catching what she is saying* correctly and seeing things correctly and translating it correctly.

Me: And you can tell me if this is typical or not, and this is also going to be very, very challenging for a lot of folks who aren't comfortable with this whole idea. But beyond the specific data that was evidential, and there was a lot of it, there were a lot of just uncanny coincidences that happened with Megan's mom during this process.

In terms of emails that she got, in terms of lights going on and off, in a way that she had associated in the past with Megan's presence being close to her, and in terms of her just knowing and feeling Megan's presence and feeling resolution to these things. So I don't know if that's typical, but it sounds like—to me, from talking to Megan's mom—that this process that she went through, and working with you without even knowing your name or who you were, the process was very transformative for her.

Hughes: A lot of times it does happen like that. Not necessarily always like lights going on and off where there is physical phenomena, but a lot of times people will see me in dreams and I will tell them things. And it was very interesting with Megan because she was just immediate. She really wanted to communicate with her mother, and she made a lot of effort to do so.

And it was almost difficult to keep up with her because there were so many interesting twists in her story. Even with the issue of questioning her sexuality and the woman who was a lesbian, you don't want to make things worse with the family if this is something they don't know about or if it's not correct. So this was actually a very interesting case to utilize because it had so many interesting facets to it.

Me: I think we stayed very true to our original goal of keeping it blinded and scientific.

Hughes: Totally blinded.

Michelle

Immediately after revealing the results to Hughes, I called Megan's mother, Michelle, to inform her about our conversation:

Michelle: *One of the most profound facets of the information that Marilynn sent was where she conveyed that Megan was conveying to her that she had been seeking and searching spiritually before her death, and that she had been confused. She had been misled, and now she was communicating through Marilynn that she's on the right track now. She had stumbled into some wrong paths, and right before she died she was reading the Bible every day and praying, and she felt like that was where she needed to be. And I thought that was so profound that Marilynn was able to interpret that through Megan. And she knew nothing about that.*

Me: *I thought it was pretty remarkable too. It's kind of hard, objectively, to pick out that kind of information and say that that's some general information that a medium would give. And again, through the whole process, suicide was never mentioned to Marilynn. She never knew until half an hour ago when I spoke to her on the phone.*

Michelle: *Really?*

Me: *Yeah, that was never disclosed to her. And she never knew the exact conflict in the family that was surrounding the issue of Megan's passing until I just told it to her. And she was like, wow, that really makes a lot of sense, in terms of what she was getting from Megan.*

Michelle: *I can tell you, Alex, that there has been a huge healing component in all of this. And as sad and as traumatic as it has been, if there can be good, there has been good from it. And this has been huge. And it just was a reminder that we really are just "spiritual beings having a human experience." And I know that she's still here with me.*

And you did mention the lights going on and off, and yeah, that has been a very, very real phenomenon ever since she died. And more so in the beginning, but it did happen the other day when I was specifically communicating over the phone to my husband. I was in tears, and I had just received this email, this answer that I had been looking for all this time, and I was very emotional. And my husband said—he had read the email from Marilynn—and he said, "Marilynn said that Megan said that she is going through this process of healing and understanding where she is, which is different from where you are. But it's still a process. And she's going through it just like you are, trying to understand and seeking answers." And when he said that and I started crying the electricity in the entire house went out. I've lived here two years and that has never happened.

When she died, I lived in an apartment and it used to happen all the time. But it had not happened here. And for it to happen at that moment—that was confirmation that Megan was listening in on that conversation. Megan knew exactly what we were talking about and that was her way of communicating, "Right on. That's exactly right."

Marilynn Hughes' reading was nothing short of remarkable. Not only was much of her information specific and unexplainable, but the deeper, transformative aspect of the reading was impossible to ignore. Does this mean we should believe everything Hughes has to tell us about the "spiritual realm"? I don't know, but after sitting in on many readings and having several of my own, I know they somehow allow us to connect with the deceased. Science's unwillingness to explore this powerful, grief-relieving phenomenon is shameful but not unexpected. As we've seen, science-as-we-know-it has a profound fear of death.

CHAPTER 7:
WRONG ABOUT TELEPATHY...
DOGS THAT KNOW...

There is a kind of materialist ethos in science...when cosmologists come up with the idea that there are multiple universes, billions of actual universes besides our own for which there's no evidence at all, instead of this causing outrage, it becomes totally mainstream... no one suffers an attack as a result...The reason that gets past the filters is it doesn't overturn a particular ideology. What's at stake is not science itself but ideology.[26] — Biologist Rupert Sheldrake

I get a charge out of talking to really smart people. It gives me the sense I'm part of a grand expanding world of knowledge. I've met many amazingly intelligent, paradigm-busting people on *Skeptiko*, but no one has impressed me more than former Cambridge biologist Rupert Sheldrake (*Skeptiko* #1, #46, #84, #97, #134, #184, #207).

To some, he's a heretic and a maverick biologist, but I've found his approach to be simple and straightforward. Sheldrake is a *follow the data, wherever it leads* kind of researcher. To this end, Sheldrake has often devised tantalizingly simple experiments with huge implications. One such experiment was featured in his book *Dogs*

That Know When Their Owners Are Coming Home.[27] This research explored the fact that about half of all dog owners report that their dogs seem to know when they are coming home. Sheldrake investigated the phenomenon for several years. After ruling out obvious signals like noise and smell, as well as regular patterns, Sheldrake found himself facing a genuine mystery—are some dogs psychic?

The mainstream scientists I interviewed in 2007 kept telling me that replications were the only thing holding them back from accepting extraordinary paradigm changing research like Sheldrake's simple "Dogs That Know" experiment. They told me mainstream science craved new discoveries and would be delighted to jump on board once experiments like these were verified and replicated.

Naïvely, I believed them. As a result, I started *OpenSource-Science.net* in 2008. The goal was to join skeptical scientists and new-paradigm researchers in close-to-peer-review research demonstrations. These demonstrations would bridge the gap between skeptical mainstream scientists and psi researchers investigating telepathy, psychic mediums, and other such phenomena. The plan was well received, and even skeptics like Dr. Michael Shermer lent their support to the idea.

At the time, I was not only naïve about science's interest in parapsychology, I was also naïve about the state of the parapsychology field in general. Later, investigative journalist and author Steve Volk (*Skeptiko* #141), who had done extensive research into this topic for his book *Fringe-ology,*[28] revealed to me how parapsychology had been co-opted and marginalized by a science-as-we-know-it establishment that sought to reinterpret ESP, telepathy, and other psi phenomena within a status-quo science worldview.

My Replication of Dogs That Know

Open Source Science's first undertaking was a replication of Dr. Sheldrake's experiment on the telepathic link between people and the animals they love. Although Sheldrake's 1997 book, *Dogs*

That Know When Their Owners Are Coming Home,[29] had detailed highly significant, peer-reviewed results from hundreds of trials, mainstream scientists ridiculed him. This made Sheldrake's work a perfect candidate for an Open Source Science demonstration.

The "Dogs That Know" project turned out to be a major undertaking. I interviewed animal behavior experts, dog trainers, and several psi researchers investigating telepathy in humans. I ran advertisements for dog owners witnessing this behavior in their animals. I even purchased laptops and cameras and sent them to dog owners so they could record their pet. Eventually, the effort paid off and we achieved promising results. In April of 2008, I published a YouTube video of one dog demonstrating the same kind of psychic-waiting Sheldrake found in his experiments.[30]

Next, I contacted Dr. Clive Wynne at the University of Florida's Canine Cognition and Behavior Lab. After interviewing him on *Skeptiko* (*Skeptiko* #47) I decided to award him a small research grant to investigate the human-animal bond and its implications for human consciousness. Although Dr. Wynne was skeptical, he was willing to give it a try:

Dr. Clive Wynne: *I am thrilled to be involved with what I hope will be the definitive study into how dogs know when their owners are coming home. This is important work that will deepen our understanding of the dog-human bond.*[31]

Despite my high hopes, research at the University of Florida never got off the ground. The dogs cooperated, but the people did not. The project became saddled with technical and logistical problems. After a few short months, Wynne threw in the towel. I was dumbfounded. Here was a simple experiment with the potential to radically alter our understanding of human consciousness, but he didn't seem interested.

While I was going about my own experiments with "Dogs That Know," I was also digging into Sheldrake's original experiments. Were they solid? Had he made mistakes collecting and analyzing the data? I found one serious scientist who seemed to think so. Popular British psychologist Dr. Richard Wiseman had built a reputation debunking all manner of parapsychology claims. He took on Sheldrake's research in a paper he published in the *British Journal of Psychology* in 1998.[32]

Rupert Sheldrake

I managed to interview Dr. Wiseman several times (*Skeptiko* #11, #35, and #97) about his controversial collaboration with Sheldrake and his analysis of the Dogs That Know experiment. Most *Skeptiko* listeners found Wiseman's distortion of the experimental results reprehensible. Moreover, the evidence in favor of Sheldrake's interpretation of the events was so clear that the case became something of a litmus test for me. It became a means of separating the curiously skeptical from the dogmatically skeptical. Finally, the controversy was brought to a head with the publication of Dr. Richard Wiseman's book *Paranormality*[33] in 2011, in which Wiseman not only misrepresented the results of the experiment, but inserted himself in the center of Sheldrake's research.

I interviewed Sheldrake in April of 2011 (*Skeptiko* #134) and asked him about his research and Wiseman's claims.

Dr. Rupert Sheldrake: *Well, the research was designed to look at the idea that telepathy and other psychic abilities are part of biological nature. If dogs have them, we have them, not because we're special or supernatural but because they're natural and these are normal abilities which occur in animals. So as a biologist I thought, "Well, if these phenomena exist at all, they should exist in animals."*

That's where I looked for them, and I collected more than 5,000 case histories from people who work with animals, including dog trainers, police dog handlers, blind people with guide dogs, riders, etc., and also vets and cat owners. I found that many people had noticed things that suggested that animals could pick up on their owner's intentions.

One of the easiest to test scientifically was the claim that many people make that their dog or cat knows when they're coming home and goes and waits at the door or window. The people see it waiting there and they know when the absent person is on the way. I found that there had been virtually no research into this. Skeptics dismissed it as being a natural routine or the dog picking up sounds from the person coming home like a familiar car engine or smells from miles away or whatever. They explained it away.

But I did proper experiments to test this. We had people come at random times in unfamiliar vehicles and we filmed the place where the dog waited. We found that with a dog called Jaytee, 85% of the occasions when the owner was coming home, the dog was indeed waiting for her. He started when she decided to come home, before she even got into the vehicle, and he waited there most of the time. This happened at random times of day in different vehicles, taxis, and other vehicles that she'd never been in before.

We built up a body of evidence from this and other dogs showing that there seems to be a real ability and that it seems to be a matter of the dog picking up the person's intention. So that's really the background to this.

Me: What intrigued me all along was that you devised a brilliantly simple experiment that can pull anyone in who's ever had an encounter with a dog or an animal and can immediately understand what you're trying to do. You really did a great job with this and I was immediately interested in your research. But at the same time, what hooked me was the reaction you got from that side of the scientific community that really reacts so strongly when anyone suggests the underlying assumptions we have about human consciousness might not be exactly the way they assumed.

In this case, Richard Wiseman was that guy to take on the challenge, so if you could go back and tell us a little bit about the first round with Richard Wiseman and Dogs That Know. I think that *provides an interesting historical backdrop for what's happened recently.*

Dr. Sheldrake: *When I was doing this research, there was a program about it on television based on an experiment that was done by Austrian State Television showing that Jaytee really did do what I had found him to do. They filmed it with a TV crew. It was a science unit and they did a proper experiment. It worked dramatically.*

Jaytee didn't go to the window when [Pam] wasn't on the way home, and he waited there when she set off to come home at a random time, even before she got in the taxi. Now, skeptics were asked to comment on this, and Richard Wiseman was, and is, one of Britain's leading media skeptics. He said it was just a matter of routine or it must have been familiar car sounds. He made all the standard criticisms, which were things I'd thought of in the first five minutes that I'd taken an interest in the subject, which we'd already ruled out.

So I invited him to do some tests of his own, which he did, and they gave essentially the same results. In my tests the dog was at the window about 4% of the time when Pam wasn't coming home, taking all the time she was out apart from just before she went home. When she was on the way home, in my own randomized experiments, the dog was at the window about 55% of the time, on average. In Wiseman's tests, the dog was at the window 4% of the time when she wasn't coming home—the same as my own—and 78% of the time when she was coming home. This shows an even bigger effect than I'd observed.

Incredibly, he then appeared on TV and made press releases, [and] wrote a scientific paper in a scientific journal claiming to have refuted the dog's ability by saying he'd given a false alarm before she set off to come home and therefore the rest of the data could be discounted.

He then produced a video clip which he's just put out again on the internet as part of his book launch, showing the dog going to the window, and it makes it look as if the dog just goes over and over again and there's no pattern or reason and it's just coincidence that it's waiting at the door when Pam's coming home. It's deeply deceptive.

His own data shows, as you've shown by discussing them with him on Skeptiko, that, in fact, he replicated my own results. But you wouldn't guess any of that—you wouldn't know that he admitted this on Skeptiko in his new book. The new book, Paranormality, which is just being launched here in Britain, starts with the very opening sentence as:

"As I gazed deep into the eyes of Jaytee several thoughts pass through my mind. Was this cute little terrier really psychic? If not, how did he manage to make headlines around the world? At that precise moment, Jaytee gave a small cough. He leaned forward and vomited on my shoes..."

Me: Well, we have to give him some credit for his writing style, which he's always had...

Dr. Sheldrake: Oh yes, he's a good writer. This book is excellent light entertainment.

Me: It's just a deception.

Dr. Sheldrake: The whole book is a deception. Richard Wiseman started his career as an illusionist, a stage magician. His Ph.D. was on the psychology of deception. In this book he suddenly practices what he preaches. What he tells us is that people see what they want to see, and he instructs his readers in the arts of deception so they can fool their friends and carry out cold readings and pretend to have psychic powers. Interestingly, he even shows people how they can airbrush away the past.

Me: Let's talk a little bit about that. Whenever you say someone is deceptive or someone is airbrushing history, it sounds so horrible that people immediately dismiss it and say, "Hey, you must have a grudge. You must have something against Richard Wiseman." But anyone who looks into this case, the deception is so obvious. The misinformation is so outrageously obvious that it's just hard to understand how he can even think he could get away with it.

But then again, and now we get into the interesting part that we'll talk about later—and that is, he knows he's going to get away with it. He already has gotten away with it for a number of years. If we look at it historically, the history of psi, and the history of parapsychology research, [critics] have gotten away with [this kind of misinformation] over and over again. So there's no reason to think that he wouldn't be emboldened in this effort... because it's built on a history of this kind of misinformation.

But before I get too far with that tirade, I'd like to break down, in very simple terms, some of the misinformation that we have here. What I'd direct your attention to is his representation of how he first came to know Pam and Jaytee, because here's a case where you talk about airbrushing someone out of history, you are not even mentioned as being a part of this experiment with this dog, Jaytee and with her owner, Pam Smart. I mean, you're not even mentioned. This is all Richard Wiseman if you read his book, correct?

Dr. Sheldrake: Yes. I come in as an endnote. He says that I also did some experiments with Jaytee and came to different conclusions. That's the only mention. Most people don't read the endnotes. And it sounds as if [he was responding to] a preposterous claim on television about a psychic dog. He investigates it. He then claims that he's shown that it doesn't really exist and that's the end of the matter.

He doesn't mention the fact that he did four tests altogether. I did 200. I'd been doing it for a year before he appeared on the scene, and the only reason there was a TV show at all was because I'd organized an experiment to test this for the science unit of Austrian State Television.

Me: Correct me if I'm wrong—he doesn't mention that he used your camera!

Dr. Sheldrake: I know. The whole thing was entirely parasitic on my research. Then he portrays himself as this kind of heroic *debunking figure that goes in and exposes people who fool themselves about their dogs and so forth.*

It is completely outrageous, but as you say, he's got away with it before. He's now creating the illusion that the whole thing's been dismissed and that's the end of psychic pets. It's just a matter of people fooling themselves.

Me: *What is going on here? What do you think is really behind this? What's your best guess, having been in this for as long as you have?*

Dr. Sheldrake: *I think it's a tendency for people to see what they want to believe, to believe what they want to believe, to only notice evidence that fits their dogmatic point of view or their belief system. He himself is a perfect example of that. He accuses people who are interested in psychic phenomena and do research on them in an open-minded way of being fooled or of self-deception.*

Basically, Wiseman is a dogmatic materialist. People who are materialists aren't people who don't believe anything; they're people who have a really strong belief that the mind is nothing but the brain, that free will doesn't really exist and we are just robots. He tries to prove in this book that we are deceived by our senses into believing things that fit with our pre-existing beliefs.

What this is, is something that pretends to be scientific but which is not. And the word for that is pseudoscience. What we're dealing with here is a particularly virulent form of pseudoscience. It's propelled by this very deep belief system and because they're so sure they're right, so convinced that the materialist worldview has to be true and that it's equivalent to science and reason, then there's not really much point in wasting time on evidence you know in advance must be false.

In 2010 I ended Open Source Science's Dogs That Know project. While it was frustrating to give up on a project that had shown such promise, I did gain a better appreciation of the obstacles involved in gaining scientific acceptance for evidence that defies science's orthodoxy. But, as you'll see in the next chapter, I wasn't prepared to accept how far some will go to defend it.

CHAPTER 8:
WRONG ABOUT PSYCHIC DETECTIVES...

[T]he scientist makes use of a whole arsenal of concepts which he imbibed practically with his mother's milk; and seldom if ever is he aware of the eternally problematic character of his concepts. He uses this conceptual material, or, speaking more exactly, these conceptual tools of thought, as something obviously, immutably given; something having an objective value of truth which is hardly ever, and in any case not seriously, to be doubted...in the interests of science it is necessary over and over again to engage in the critique of these fundamental concepts, in order that we may not unconsciously be ruled by them.[34] — Albert Einstein

    In 2004, Court TV aired a series entitled *Psychic Detective*. It featured interviews with homicide detectives and other police officials who had used psychics to help solve cases. It was a compelling show. One episode that caught my attention was the case of New Jersey serial killer James Koedatich. The story was remarkable in a number of ways, including the unbelievable resolution in which the killer essentially turns himself over to the police after experiencing a psychic attack.

    In this chapter, you'll follow my investigation into psychic detectives and how it led to a fundamental shift in my understanding of why some science advocates believe what they do.

Beginning in September of 2008, I participated in a joint investigation of a psychic detective case. My partner in that investigation was Ben Radford, a science journalist and self-described "skeptic." To some, those who write for the *Skeptical Inquirer*, or Michael Shermer's *Skeptic* magazine, can be easily dismissed as biased. Readers know what they're getting. But, like many high profile skeptics, Radford is "mainstream friendly" and regularly writes for *Discovery News* and other mainstream science publications. His influence stretches beyond the fringy skeptical community.

Moreover, and most importantly, Radford's dogmatically skeptical position regarding psychic detectives and psychic/medium communication is completely consistent with mainstream science. In this way, people like Radford serve as tip-of-the-spear advocates for maintaining the status quo at all costs. They provide the cover science needs to go on ignoring these extraordinary accounts.

As you'll see, the investigation was quite extensive. It spanned months of work and included multiple transcribed interviews with all the key players. The conclusion was self-evident—the police detectives repeatedly corroborated psychic detective Nancy Weber's amazing account. They may not know how she did it, but they know that what she told them matched the facts of the case.

Amazingly, Radford still denies this fact. Even more astounding, Radford has publicly referenced this case as an example of how to successfully "debunk" paranormal claims.

Ben Radford

In November of 2008, I published a *Skeptiko* episode titled, "The Psychic Detective Challenge" (*Skeptiko* #57). During the show I asked Radford, "Why don't you ever investigate good cases?" He replied, "Why is whatever case we pick never good enough?" Here is an excerpt:

Me: *...I've read many of your articles and you're obviously a very talented writer; you do a good job of summing up these cases. You're not flying off the handle in a really incendiary way, but I do feel like a lot of times you pick on cases that have failed and make the case that this somehow proves something. I was reading and I was going, "Gosh, why hasn't he looked at the best cases?"*

Let me give you an example. Here's someone who I've actually spoken with, a psychic I've spoken with. Nancy Orlen Weber is her name. She was featured on the television show Psychic Detectives. *I went to her website in preparation for this interview. I've seen your reports, Ben, where you've said these psychics appearing on TV never solve any cases. So, you go right to her website and you read her letters of reference from detectives, detectives who've been in the New Jersey State Police for 25 years, and they say: "Yes, I've worked with her on all these cases and she really helped; Yes, I spoke with these other investigators who worked on murder cases with her; she did a really good job." She received a letter of commendation.*

I mean, there's like this total disconnect. Do we really care that there are all these failures, or do we care that the phenomenon does seem to happen sometimes with some people to the extent that it's proven beyond the doubt of the folks involved—the New Jersey Police? These are people you're not going to dupe with some tall story. Where's the investigation on the best cases?

Ben Radford: *Why don't we do this; you find the best case you can find. Look through all the psychics you want, figure out one, and pick the one case you think is airtight. Give it to me and I'll get back to you in a couple of months, and we'll see what we'll find.*

This became our "Psychic Detective Challenge." I'd find a "best case" of psychic detective work and we'd jointly investigate. However, before starting, we faced another challenge: we disagreed over the criteria for evaluating success. I felt the police should be the ultimate judges. If they felt the information given was useful, then it was useful. Radford insisted only a qualified "skeptical investigator," such as him, could determine whether the information provided was remarkable and/or useful. We decided to go forward and let the *Skeptiko* audience decide.

Tasked with picking a "best" psychic detective case, I called up Nancy Orlen Weber and explained the project.

Me: *What I've been doing for the last couple of years is to try and level the playing field a little bit in the way hardcore skeptical science reporters treat these topics. Here's what I'm focused on right now.*

I've recently had this encounter with the editor from the Skeptical Inquirer *magazine. We've had a nice exchange, a pleasant, cordial exchange, but I've challenged him on his skeptical investigation of a psychic detective case. So he said, "Well, why don't you pick a case? You pick a case and we'll investigate it together."*

Nancy, I don't know if you run into these people, but he maintains that there is not one single successful psychic detective case in history. So, would you be willing to help me go through a case that we can find, that's reasonably well documented, where there's law enforcement that would at least spend a few minutes confirming the information that's already been out there?

Weber agreed to help and suggested we focus on the Amy Hoffman murder case.

In 1982 Amy Hoffman, a high school cheerleader, was kidnapped from a Morristown shopping mall. Two days later she was found stabbed to death. Police had no suspects when twelve days later 29-year-old Deirdre O'Brian was abducted from her car, raped, stabbed, and left for dead. She died soon after. On the night of January 16, Weber held a prayer group meeting and asked that the perpetrator of these crimes experience some of what he had inflicted on the victims. Later that night James Koedatich phoned police complaining he had been stopped by an assailant and stabbed in the back while driving through Morristown. Upon investigation police found no stab wounds on Koedatich but noted that his tires matched tracks discovered at the murder scene. Further investigation turned up sufficient evidence to arrest and convict Koedatich.

This case was well known to those in Pennsylvania and New Jersey, but until the Biography channel did a show featuring Weber's work the case was not nationally known. Fortunately, Weber had remained in contact with the detectives who worked on the case and was able to provide what we needed to get started. I called Radford to tell him the news:

Radford: Sure, absolutely. Let me just say at the onset, before you sent that name, I had never heard of it. I look forward to looking into it. If it really is a good example of a psychic detective, then great.

So we finally had our case. Next, we needed to investigate. I had spoken with Nancy Weber about the investigation and came away impressed. Now, I wanted Radford to talk to Weber. We scheduled the conversation, but before the conference call I had an interesting chat with Radford about the investigation (*Skeptiko* #69):

Radford: So I think we can all agree on what makes this case remarkable... is the accuracy of the information. The Weber case, I mean. You know, usually, and I [don't] need to tell you this, but usually, you know, a lot of the psychic detective information is general...the body will be found near water. Well, is that a pond, is that a lake, or is that a stream, or is that a ditch?

Me: *Yeah, I do not think that is really the case. Cops have no patience for that. If the information is not useful, they are not going to use it. This idea that cops are going to be duped by someone saying, "... it's going to be found near water..." I don't think so.*

Radford: *Okay.*

Me: *But I want to point out that what Nancy finds remarkable about this case, and what Captain Moore and Lieutenant Hughes find remarkable [about] this case, is the ultimate way the case [was] resolved...that it's Nancy's prayer that then leads to Koedatich's ap-*

prehension. But that would be hard to establish in a concrete way.

Radford: Right, right, yeah, yeah. You know...you are exactly right...in a case like this, there [are] so many elements to it, there [are] so many facets to it. And we said, "okay," ...regardless of how he was eventually caught ... what [we] are going to focus on was the specific information that she gave to the police regarding Hoffman's killer before they knew it.

Me: Right.

Radford: So anyway... when I am looking at this...I think it is important to realize that in this case, the burden of proof is not on me to disprove it.

Me: Of course it is.

Radford: No, it is not, the burden of proof...it is just like in our court system. The burden of proof is on the person making the claim.

Me: No. If it was going to trial, you would present your evidence and I would present my evidence, and then a jury would go in a dark room and decide based on the evidence.

Radford: Well, that is fine...as long as we agree that, you know... at least in the beginning... it is not up to me to prove that... if someone says that a UFO landed on his lawn, and [he] tells me that, it is not my job as a skeptic investigator to prove that a UFO did not land on his lawn. It is up to him to prove that it did. So, I mean, that is an important distinction to make. But otherwise...

Me: Right, but let us just be clear. I mean, in this case, it is not quite like [that]. In this case, there was a murder.

Radford: Right.

Me: There was someone apprehended.

Radford: Right. I will tell you what I will do... from my point of view... the information [that] has been given [thus far]... I do not see a whole lot of evidence being presented to me other than the testimony. But, I am willing to take on the burden of proof...

Me: Well, if you can't rely on two cops who did not know each other before the case... where one is a Captain and has run dozens—if not hundreds—of investigations like this... if you cannot rely on their testimony for even the most basic facts of a case... well, then, the burden of proof that you have in your mind is way, way different than the burden of proof that I have.

Soon after this brief exchange, Weber herself joined us on the line:

Weber: Hello!

Me: Hi, Nancy! It's Alex Tsakiris and I have Ben Radford from the Skeptical Inquirer on the phone as well...Nancy, thanks again for taking the time to do this. And I really want to say...I really think Ben has really done a very good and admirable job coming at this thing from a skeptical perspective...he has really dug into it a lot. I have dug into it a lot [as well], and we [have] both spoken with Capt. Moore and Lt. Hughes, but since Ben [hadn't] had a chance to talk you... we thought we'd get you on the phone and just take one more crack at this because I think... so many times in these kinds of cases, no matter how much time you spend, there's always [some] unanswered questions.

Weber: Absolutely!

Me: And it is really hard for some people to accept the idea that you are somehow, through some anomalous means that we do not totally understand, getting information and giving it to police, and they are actually using it in investigations. That is just really hard for a lot of people to swallow. And I think Ben is skeptical, but I think he is coming [at it] from an honest, open, "show me the data" kind of [perspective].

Weber: Well, I feel the same way about light switches. I do not fully understand electricity.

Me: (laughing)

Weber: But if the result is useful, I am all for it...

Radford: Right.

Me: ...That's a good point because one of the things Ben and I were chatting about before is that this process is almost like a legal *proceeding where each side is presenting their case...*

Weber: *Right.*

Me: *And you have mentioned in the past that...you have worked on over 300 cases with law enforcement.*

Weber: *Probably... I do not know how many cases I have worked on because...*

Me: *More than a hundred?*

Weber: *Oh, yeah.*

Me: *More than two hundred?*

Weber: *Probably more than [300], [close to 400], [but] I really do not know because I do not keep records of most of them.*

Me: *And have you worked with a lot of different police forces?*

Weber: *Yes.*

Me: *And do any of them come back to you again and say... "Help [us] with another case"?*

Weber: *Yeah! Some of them do, some of them don't. It depends on their status and their community of law enforcement...sometimes I have to go [through] an awful lot just to get accepted...*

Me: *Right.*

Weber: *But then, for instance, I worked on a case with the state prosecutor's office... We spent the day and we worked on it. [Most times they] don't feel it's appropriate to let me know what's going on with the case; a lot of them. So, I never find out. But then, I get a call from somebody else from that state, from another department, because I was referred, which means that they liked what I did. Now, do I know what they like? No. Would they tell me? Doubtful, because they do not want it to enter their case in court...*

Me: *Go ahead Ben... why don't you jump in.*

Radford: *Yeah. I was going to say again, thanks for taking the time to talk on this and [share] your story...I have been looking into, you know, psychic detectives and psychics for a long time and one of the common things that you can find, and even you might admit this, but often times [the information provided] is very general...*

Weber: Yeah, sure.

Radford: The body will be found near water. Well, is that a lake, is that a...

Weber: Yeah, the whole planet is a body of water (laughing)...

Radford: And what specifically in a nutshell... did you [tell] Moore and Hughes about Hoffman's killer?

Weber: Okay, sure. I saw a tall, a fairly tall maybe 5'10"-5'11", slim-built man, apple kind-of shaped face. And I saw a lot of dark hair on his head. And then I saw the hair being taken off almost like a wig. And it would be a buzz-cut with a very high widow's [peak]. I kept seeing that.

And then I got the name "James" as I was writing it. And then I got that he knew the Hollow in Morristown, which is a particular area in Morristown. As we were riding around, I also smelled gasoline, smelled an auto shop or a gasoline station. And I knew he was associated with one, and then I got the word "brother." So to me, his brother had either owned or worked at the station, and the car that James would be driving would have that [same smell].

And then I got Polish, and I connected that as Polish descent. That took me to Florida for some reason, where I felt imprisoned, like I was locked away in prison for murder. And I remember saying to them, "Oh my God! He was in Florida in jail for murder, not women. And they let him out."

And you know, as I'm saying it, I'm getting it. I do not know if that makes any sense to anybody, but I did not know it and then said it. I just let the words spill out, "They released him, those idiots."

Radford: Okay.

Weber: I said I hope...the family sues the state for false release of a man who [is] a mass murderer. Crazy. And that is how it was coming. So, then I got the last name was Polish, that began with the K and ended in "ish," and I could not get it. It was K-blah-blah-ish. And that is what I had said.

Then I felt that he left something of Amy's behind in his car. And that...it would still be there. And they said "Oh, doubtful." I said "No. It will still be there." And then [we] went to...where she was murdered. They were driving around and I kind of directed them down there to the road because I just felt like something like a rope was pulling me. And then when I walked, they walked behind me. And I walked over to an area and I got down on the ground, and I thought I would just fall apart, because I could feel her pain. It was very fresh.

Radford: *Like you are channeling her spirit?*

Weber: *Well...nobody really knows what it is, but for a lack of any other understanding of it, yeah; she was sharing with me everything that happened, and I was re-experiencing it for them. And so I went [through] the whole thing. And I told them what I was feeling and seeing, and that he had a knife, and the knife was the main weapon that he used...for [everything] including...getting her into the car.*

Then afterwards, we drove around...and as we went [further] I said "Take me to the police station. I do not know where it is, but hurry up, get me there." [Laughing.] So we went into the police station. And they introduced me to the captain.

They [told him], "Please listen to her. No matter what it sounds like, please hear her." And I said, "You have a man who gave a ticket, and maybe two speeding tickets, to a man named James, beginning with the last name K, who grew up in Morristown, Polish decent. And the officer who gave it to him has a last name that begins with "C."

The Captain said, "We have two." I said, "The one with the hard 'C,' not the soft 'C.'" So he said, "That would be Constanza." I said, "Yes, that would be it, and he knows who I mean, and that is your killer. Please talk to him." Oh, what I did not know at the time [was] that they did talk to Tommy Constanza about [a man named] Koedatich, and they called him in to the prosecutor's office. In the prosecutor's office, a person who I won't mention said, "Oh, we

looked at him...and let him go."
Then later:
Radford: *Right...part of the problem...comes down to memories from 27 years ago. So if Moore and Hughes would recall differently, then either of them is remembering [incorrectly] or mistaken, right?*
Weber: *If they recall...what?*
Radford: *Well, I mean...recall this specific information that you gave...*
Me: *Where are you going with this, Ben? I mean, I talked to Moore and Hughes and they said the exact same thing that Nancy said. How was it any different?*
Radford: *Well, in fact, because I interviewed both Moore and Hughes. And like for example with "Florida." You said that he came up from Florida, and he had been imprisoned there for murder.*
Weber: *Right.*
Radford: *But that is not what Moore and Hughes told me.*
Me: *Yes, it is. That's what they told me. I have it recorded.*
Radford: *Right, and so do I. I mean I interviewed both of them, I have a recording of it, Alex has a copy of it; we can always go back and ask them. Sgt. Hughes told me, and I quote, "I do not remember... I do not recall her specifically saying he done time in Florida, just that he had done time in the South."*

We Speak to the Homicide Detectives

I know it may seem bizarre for an intelligent journalist like Radford to be fixated on such a minor point. Weber had provided an extensive, incredible reading to the detectives and he is hung up on whether she specifically mentioned "Florida," or just "the South." But unfortunately this is where the joint investigation had led. It only got worse once Nancy left the call (*Skeptiko* #69):
Me: *It is nonsense, Ben. I have Jim's transcript, which you could read yourself...*
Radford: *I have read it.*

Me: OK, here is the transcript of my interview with Captain Jim Moore, "...she also told me that the individual came from the Morristown area. And that he had moved to Florida where he had committed the murder and was sentenced to jail. While in prison, he committed the second murder of an inmate. And he eventually got out, and returned to the Morristown area." [Again] this is Captain Jim Moore talking, "that either him or a member of his family or gas station... she told me that he was very upset with the man at the police department..."

So all those things directly corroborate Nancy's testimony. If you just stop right there, you would say that's remarkable. There was no way to explain how Nancy could come up with that information, and how Moore could corroborate it.

Radford: But you have three different people who can't get this thing straight... Alex, her testimony is, quote, "He came from Florida."

Me: And that is how Jim Moore recalls it.

Radford: That is not. No, no, no, let us be very crystal clear. Jim Moore said that [Nancy] said that he did time in the South.

There was only one way to resolve this disagreement. I decided to conference Captain Jim Moore into our conversation and ask him directly:

Me: Ben and I had an interesting chat with Nancy this morning. And we got off the phone and we started arguing like we do on Skeptiko *[laughing]*...

Radford: (laughs) We do that.

Me: But you know Ben has some potential inconsistencies he sees between the account that you gave to me and the account that you gave to him. We thought the easiest way to clear that up was to get to you on the phone for just a few minutes. I'm going to let Ben to take over. Go ahead, Ben.

Radford: Yeah, I was just going to say, you know, part of the problem in this case is that, of course, you know, there [are] no notes.... You do not have them, Hughes [does] not have them, Nancy does not have them as far as we know. So all that we're going on are people's memory and, of course, you know, detectives know memories change over time. So I am just trying to figure out where the truth stands.

So for example, Nancy claims that she specified that Koedatich, [you] did not know [the name] Koedatich at that time, but Hoffman's killer comes from Florida. But both you and Hughes said that you recall that it was not necessarily Florida, it was just in the South. Do you remember which one of [those]?

Capt. Moore: No, no. I said Florida. She told me that...he lived in the Morristown area, went to Florida. While he was in Florida, he killed someone, went to jail. When he was in jail, he killed an inmate. That is what I told you.

Me: And that is what you told me on my interview and I have a transcript of that as well.

Capt. Moore: Yeah.

Radford: Ok. So what [Lt.] Hughes told me is...I quote, "I do not remember her specifically saying he done time in Florida just that he'd done time in the South."

Capt. Moore: No, I did not say that.

Radford: No, no, I just said that [Lt.]Hughes said that. So I'm just trying to figure out why. I mean, if she told both of you the same thing, I just find it odd that he remembers something different but...

Me: But he does not remember different, Ben...

Capt. Moore: No, no, no. I specifically told you he was in Florida, and he did jail time in Florida, and he killed an inmate while he was in prison in Florida.

Radford: No, no, no, I understand that is Koedatich's history. But the question is: is Koedatich's history the exact same thing that Nancy Weber told you?

Capt. Moore: If you want to know what Nancy said to me...
Radford: Yes.

Capt. Moore: *Before Koedatich was arrested, she told me that he, whoever committed this crime, went to Florida. [Before] leaving Florida [he] killed someone, went to prison in Florida, and killed a [fellow] inmate while he was in Florida. For some reason, unknown, he got out. And after he got out of prison, he came back to the Morristown area.*

Radford: *Okay.*

So, you might think we had resolved the issue, right? Captain Moore could not have been more clear about what he remembered and what he told Radford. If you listen to the actual recorded interview (*Skeptiko* #58, and #69) you'll hear just how annoyed this veteran homicide detective was getting with a reporter who keep insisting he had said something he didn't say. It made me cringe. But, this wasn't the end of this controversy. Radford might have struck out with Moore, but he was still insisting this issue, more than all the other remarkable information Weber had provided and the detectives had verified, was key to his skeptical investigation. He shifted his focus to Lt. Hughes suggesting he would back up his version of events. We conferenced in Lt. Hughes.

Me: *... I am not going to take too much of your time. Let us turn it over to Ben. Ben is concerned about some potential inconsistencies he sees between the interview you gave with me and interview you gave with him. So let us let him kind of take it over.*

Radford: *..Actually, I'm trying to reconcile what Nancy said. This is all regarding the information she gave in '82 before Koedatich was arrested. For example, she says that she said specifically that the Hoffman killer came up from Florida, but from what you told me...you said that you did not remember her saying he came from "Florida" just in "the South." Is that right?*

Lt. Hughes: *No, no, no. I might have said that, but no, she knew that the guy had killed before and that he had killed in Florida.*

Radford: *Ok.*

Lt. Hughes: *Yes. She knew that he had killed before, that he had* killed in Florida.

Radford: *Ok ...*

Lt. Hughes: *That he had done prison time in Florida.*

Radford: *Ok, you kind of told me that. You said, "Done time in the South." You said, "I do not recall her specifically saying he done time in Florida, just that he done time in the South."*

Lt. Hughes: *No, I do remember Florida. Maybe I was being generic or whatever the case might be, but I do know that she said Florida.*

Just Ignore the Facts

Whew! Glad that was resolved. All the principle players in this investigation agree psychic Nancy Weber gave an impressive reading. All agree she was inexplicably able to provide relevant information about the perpetrator, including the fact he had served time in a Florida prison for murder and had been accidently released.

But wait...there's more to this story. As I mentioned at the beginning of this chapter, Radford not only refused to acknowledge what his investigation revealed, but in a way I've grown accustomed to seeing with dogmatic science-as-we-know-it defenders, he portrayed it as a successful debunking for his audience of disbelievers.

Skeptic or not, this kind of careless disregard for the facts by a science journalist was stunning. And it turned out to be somewhat of a turning point for me. I had spent months jointly investigating the Amy Hoffman case with Radford. I had worked closely with someone I thought was an honest, fair-minded colleague; someone who, like me, was willing to follow the data wherever it might lead. But I came away with a new understanding of how far some otherwise rational people will go to protect their worldview. I also caught a glimpse into how this all-too-common human weakness can undermine science's search for truth. After Radford published his "findings," including the "inconsistency" between Florida versus "the South," I produced a final *Skeptiko* episode on the topic (*Skeptiko*

#151), which I concluded with these words:

Me: *So I don't know what you want to call that—misrepresentation? Misremembering? Uh, lying? I don't know what kind of words you want to throw at it, but it's stunning. It's stunning but it would be more stunning if it hasn't happened over and over and over again on this show. We have countless examples. All you have to do is go through the archives and listen to Richard Wiseman and listen to Ray Hyman, and Steven Novella, and James Randi. I mean, it's a consistent pattern. We can't even call Ben out for being especially over-the-top in terms of misreporting this stuff. It has to do with the bias, the worldview. It clouds their vision and they—just like the fanatical fundamentalist religious folks they so despise—can't get past the obvious problems in their logic. It's the same situation repeated over and over again...*

It's not always easy to ferret out the truth. But in this case, I think it's pretty darn easy. I mean, if you can't jump over this hurdle, if you can't look at what Ben's saying and say, "That's a distortion of the truth," then I don't know if there's much hope for a dialogue with a skeptic [like Ben] and a believer like me.

CHAPTER 9:
WRONG ABOUT HEALING & MEDICINE...

Science has great pretentions about itself; it basically regards itself as a meta-theory capable of passing judgment on all other theories. [These theories] are supposed to submit themselves to science to be told whether they are right or not.[35]

— Ethnobotanist Terence McKenna

Like a lot of middle-aged men, I don't like going to the doctor. But in 2013, I started experiencing severe heart palpitations. I felt like I was having a heart attack 3-4 times a day. I knew I had to see a physician. After looking up the name of the doctor who I hadn't seen in five years, I made an appointment. It's amazing how quickly they schedule your visit when you tell them you're a middle-aged man with heart pains.

After a few in-office tests and an EKG, my doctor could find nothing wrong with me and sent me home. I was relieved but still concerned. Despite the tests, I was still experiencing strong heart palpitations every day.

After a couple of months I went back to my doctor and said, "Look, I know you didn't find anything the first time, but I'm concerned and I am not generally someone who obsesses about their health." My doctor did another exam. Then she ran another EKG. Then she came back in, listened very carefully to my heart, and said, "We have got to get you right over to the cardiologist. You need to go now." My sense of satisfaction about being right about my condition quickly turned to fear about what was going to come next.

Of course, medical stories like mine often have twists and turns. With me, Obama-Care intervened. Within hours after my appointment with the cardiologist was set, I received a call back telling me Obama-Care laws had gone into effect that very day and my insurance was no longer in effect at the hospital. I called my insurance agent and found myself in insurance limbo. I would have to either change health care systems, including switching the doctors that everyone in my family had used for years, or I would have to wait 30 days for my new insurance to take effect. I decided to wait, a decision completely against my doctor's advice. In fact, she called me up and told me, "Okay, if you have a heart attack here is what to tell the ambulance." Just as I was starting to question my decision, I remembered my interview with Dr. William Bengston.

Dr. William Bengston

Dr. William Bengston is a sociology professor at St. Joseph's College in New York who stumbled upon some amazing discoveries about energy healing. He's a top-notch scientist and a stickler for methods and statistics. In fact, he's so highly regarded by his peers that he's been twice elected the president of the Society for Scientific Exploration, a group of impressive academics and researchers who apply a tight peer-review process to exploring areas of science deemed taboo by the mainstream.

After interviewing Bengston in 2008 (*Skeptiko* #185), I was so impressed by his methods and his results that I made a personal vow to give his healing techniques a try if I ever had a serious medical condition. But before I tell you the rest of the story about what happened to me, let me tell you why I was so impressed with Dr. Bengston's research and why it offers another compelling reason why we must dispense with this mind-equals-brain silliness science-as-we-know-it is mired in:

Me: *Tell us a little bit about your background, your educational background, particularly in sociology, and how that's served you. Or maybe it hasn't served you as you've tried to wrap your arms around this healing experience...*

Dr. William Bengston: *Well, I'm traditionally trained in sociology, which means you generally have a specialization or two and then it's pretty much assumed that you'll become technically proficient in designing studies and data analysis and things along those lines. I actually went a little bit farther down those lines. I'm kind of interested in advanced statistical modeling. I don't know that the sociology training, other than focusing me on becoming very, very rigorous, really helped in healing. I didn't see that as a career path...*

Me: *...But [I don't think most people would witness an energy healing and say], "Great. Well, let's run an experiment."*

...You had to have been blown away by what happened and yet your first thought was, "Where's the data? How do I produce the data? How do I run an experiment?"

...Tell folks a little bit about how you first encountered the healing phenomenon...

Dr. Bengston: *I ran into a guy who claimed that he had become spontaneously psychic. And I was skeptical. I always begin from a skeptical point of view, so I gave him a couple of things very informally to do readings on and such. I couldn't trip him up. It was very annoying. So he would do readings on personal items, wallets, and rings and things like that and tell too many detailed things. I then defaulted to, "How am I deluding myself? How can I make this effect go away?"*

So I designed a double-blind study at the hospital where patients coming in would sign a blank index card. That would be put into an opaque envelope, which would then be put into another opaque envelope, and then I would pick up the opaque envelopes. I wouldn't have any idea about their condition. I didn't even know if they were male or a female. I gave these to this newly minted psychic and he would do a physical reading on them. Then, and looking back on it now I probably broke a whole lot of laws in doing this, but I had somebody on the inside look at the medical records to see if he was accurate.

Me: Bill, what kind of reading would he give? What would he say if he has an envelope inside an envelope with an index card and just someone's name scribbled on it? What kind of information could he give you?

Dr. Bengston: Oh, he would give detailed information. He would say, "You've got a gall bladder problem." Something along those lines. So he'd say, "The lungs are filled with fluid." He would say, "There's pain in this area." Things along those lines. "There's a tumor that is located here."

It was particularly startling when I went back and got the confirmation of this. In one instance, I thought I had discovered a mistake, but when I re-checked with the hospital I found they had made the mistake.

Me: Amazing.

Dr. Bengston: Yeah. It was pretty nutty. So I couldn't make the effect go away, which was pretty annoying. And so I started to pay attention to this guy and he was doing these physical readings. This naturally morphed into something else even when the people he was doing physical readings on were claiming that the conditions were being helped on their end. And both he and I thought that was nuts.

So let's say someone has a migraine headache and he does a reading on a person who has a migraine headache. He doesn't know they have a migraine headache; he's holding something of theirs and he would get the headache. And the people would say, "At such-and-such hour my headache left." It was too coincidental. At a particular hour he would get it and it would leave them. We're scratching our heads going, "What could this mean?"

Me: You know, one of the amazing things about this whole account, this whole story of yours, is that it's so counter to New-Age-ish ideas about energy healing. Here are two guys who are saying, "What the heck's going on? Healing? What does that have to do with anything?" You're not getting any kind of divine guidance here. You're not even looking in that direction, and yet the data, as you say, is just pulling you there. So what's the next step for you in trying to make this technique operational? How did you get pulled into it?

Dr. Bengston: I got pulled into it by having myself fixed. I had had a chronically bad back for a bunch of years, and I had previously been a competitive swimmer and had to give up competitive swimming because of the bad back. As this guy, Bennett, and I are going through doing these readings and such and gathering all these silly anecdotes when it suddenly occurred to me, "You big dope, you're in pain all the time, ask this guy to help." So I thought, "Maybe if I give him something, he'll do a reading on me and maybe he can help my back."

He picked up the pain just as I thought of that. I didn't actually give him an object. I said, "Put your hands on my back," and he said, "And do what?"

And I said, "I don't know. Fix it."

And so he put his hands on my back, and quite astonishingly he did fix it. I was the first person he ever deliberately healed with hands-on.

So we went from this and said, "What do we do with this?" We started to put hands on people and watched what happened. Sometimes pain was alleviated. Sometimes it happened quickly. Sometimes it took longer. And after watching this on a couple hundred people, patterns started to emerge. For example, malignant growths responded very, very quickly. Benign growths didn't.

So if he put his hands, for example, on a wart, nothing would happen. Nothing at all. If he put his hands on a lump that turned out through medical diagnosis to be malignant, the thing would respond very, very quickly and dramatically. And so it turned out that some things he could do and some things he couldn't do, and among the things he was particularly good at were malignant tumors.

Me: Bill, let me interject here because from the little bit I know of your work you're very careful about drawing too many conclusions about what that means or what's behind that. You're more or less just seeing the patterns that exist and the personal experience. I mean, for example, your healing. You go on and you're a pretty darned competitive tennis player... so your back is obviously healed. You've got your proof.

... [But] people need to know these are experiments you've run clinically in university labs. This isn't just a guy in his kitchen doing healings. You've run some pretty rigorous experiments. Tell us about some of those...

Dr. Bengston: Starting from these case-by-case, relatively haphazard clinical observations, like malignant growths responding quickly and benign growths don't, I thought to myself, "How are we going to go from a haphazard kind of experience to very controlled conditions?" I wanted an absolute airtight test that if it worked you didn't have a viable counter-hypothesis.

Since this guy seemed to be particularly good on malignant growths, I thought, "Let's see if we can get a study with malignant growths under controlled laboratory conditions." I had a friend, Dave Krinsley, and he had an academic professorship with City University of New York. He called in a bunch of favors with the biology department and asked, "Where can we get a really airtight, no question about it, really tight experimental model?"

They were working for 20 years on a particular mammary cancer in laboratory mice, and they knew exactly what was going to happen. There were literally thousands of published studies on these mice. They get injected with a particular form of cancer, they're particularly bred—they're actually pretty inbred—and after they're injected you know exactly what's going to happen. The tumor's going to grow; it's going to be non-metastatic; it's going to kill the mouse in a certain number of days.

At the time we started this, the record [for the longest] living mouse was 27 days. No mouse in literally thousands of experiments had lived longer than 27 days after injection with this particular mammary cancer. As a researcher, you knew exactly how many mice would die and what particular day after injection because it's, again, very well documented, found in labs all over the world. So we said, "This is perfect."

The mice presumably don't believe much. We didn't believe much. We were just observing. Maybe this doesn't make any difference with the healer or the healer's beliefs. Let's find out if there's a real underlying biological mechanism going on.

So it turned out that this guy who was doing these healings, Bennett, was not particularly interested in doing experiments. I'd dragged him from lab to lab and I guess I'd pushed too many of his buttons because he backed out. Then I was left with experimental mice and no healer. That was a dilemma. It was actually borderline panic.

Since I had spent the most time with this guy, Krinsley suggested, "Well, you do it."

I thought, "Me? I'm not a healer." I've never thought of myself as someone with healing ability. I'm just someone who observed that this stuff happens when you do this other stuff. It's not my identity in any way, shape, or form.

Then again, at this point I was committed to the experiment, so I put my hands around the cages of the mice for about an hour a day. I suspected at the time that healing, if it were to work, would be something analogous to radiation. So if you take mice and you radiate them, you can kill radiated cells. I thought, "Well, if we get these mice close enough to their injection time and we zap them, as it were, with healing, then maybe something like radiation occurs and we're killing the cancer." Something along those lines.

But instead, the cancer started to grow and I thought it was failing. So the tumors grew, and I said, "Well, I gave it a shot. It didn't work and so it goes." And I said, "Let's call it off. Why put the mice through this?"

I got talked into going a little longer. The tumors kept growing, and I kept thinking, "Well, this doesn't work." But I kept treating anyway. The tumors got bigger and bigger. Then they developed this ulceration on the tumor and I really thought it wasn't working, but the ulceration grew until eventually the tumor imploded and the mice were completely cured.

Me: Unprecedented, medically... with these particular mice, right?

Dr. Bengston: Never happened before for any reason. So the world's longest living mouse with this kind of cancer was 27 days, but after it went through this process of growth then ulceration then implosion, the mice were cured. I used to say they remitted but that's the wrong word because remitted means a reduction in symptoms or temporary disappearance. These mice are cured for life. So we watched them and we leave them for two years and they would just live out their normal lifespan.

And it goes even further than that. We've re-injected them with the cancer, but the cancer won't take anymore. So the mice that are cured are cured for life.

Me: ...Let's finish this story, Bill. The world changes; you receive the Nobel Prize for Medicine. Cancer treatments around the world are revolutionized, and energy healing is the most highly researched area of medicine right now. Did I get that right?

Dr. Bengston: Except for the entire scenario—yes. It's not going to take off as you suggest. That would be my guess, yeah. So this isn't something that we've cured a bunch of mice, therefore the cancer industry folds its tent. We've got to find out what's going on.

Me: *And that's where it gets even more challenging. What do you think is going on?*

Dr. Bengston: *Well, I've replicated this experiment now about a dozen times and I use—sometimes, not always—I use sometimes other people who have no experience in healing and who are all skeptics in the real sense of the word.*

Me: *Tell us a little bit about your process in terms of screening these folks. I think it's really funny but also interesting.*

Dr. Bengston: *When I was able to cure a bunch of mice, I thought to myself, "The first thing we need to do is see if this replicates." I wanted to see if there was there some flaw in the procedure, some fluke. And I also had a lingering fear: what if I can do this and other people can't? I mean, that would be curious, but it wouldn't be very useful. So, I decided to get some people who had no experience in this, train them in the healing techniques that Ben and I had worked out, and see what happens when they do it.*

I went to a number of students and said, "I'd like you to do a healing experiment." If they responded to me, "Gee, that's wonderful, I like this stuff," or "I know about this stuff," I'd say, "Thank you for your interest," and I left. I didn't want them. I didn't want someone who knew. I didn't want someone who was already into healing.

I wanted a really blank slate and I wanted someone who—my actual criteria was they either had to laugh in my face when I told them what I wanted them to do, or they had to express extreme skepticism. I'd go to them and say, "I'm going to teach you for about six weeks on how to do some healing techniques and then you're going to get cancerous mice and you're going to put your hands around the cage for an hour a day and we're going to see what happens." And they'd go, "No, really. What do you want me to do?"

Bernadette Doran

While Dr. Bengston's unexplainable methods and too-good-to-be-true results still left a lot of unanswered questions, his approach and rigor left a serious impression on me. So, when I decided to give energy healing a try, I was determined to look for a healer that used the Bengston Method. With a little bit of help from Dr. Bengston himself, I found Bernadette Doran, an energy healer from Chicago who had received training from Bengston. I took the leap and scheduled a healing session with her. I also decided to chronicle my experiences for *Skeptiko* listeners (*Skeptiko* #252):

Me: *First of all, let me just say that this show is going to be a bit of a departure from topics that we normally talk about in terms of consciousness science, but I think the further we get into it folks will see that it is not at all a departure from the topics that we talk about. Let's start by having you tell folks a little bit about your group, E3...*

Bernadette Doran: *Equilibrium, Energy, and Education is a research-based energy therapy wellness center...*

Me: *...Give people a sense for what we are talking about when we talk about therapies and energy healing...*

Doran: *First of all, energy therapies are non-physical, or non-physical in the sense of Western medicine. All the energy therapies we do here stimulate your body's own chemistry set to rebalance itself...*

Me: *Bernadette, let me jump in there again because I can already hear in my head the pushback from a lot of people...First of all, this whole idea of energy healing is an outgrowth of this whole movement of mind-body medicine, right?*

Doran: *Yes, that is absolutely correct.*

Me: *So this whole idea that we have been exploring and that has gained some kind of credibility in conventional medicine is that there is more to the body than just the physical. So, medicine has moved toward saying, "There seems to be this interaction with the mind, and we don't know quite how it works, but heck, we have this* *placebo effect and we can't get around it and it seems stronger than our strongest medicines so maybe we better take another look at it."*

And then ... some people have gone further and said, "Well, you know, if I really dive into this mind-brain thing, we also have to look at this energy thing because even though we can't put our finger on it, this energy does seem to be able to be adjusted and shifted and when we look at other cultures we see that they seem to have a long history of doing this."

So with that, I want to say that I really support people who are willing to venture out, as you are, and say, "Okay, we don't understand everything about this, but let's start using the tools and methods of science that we have to start applying some of these techniques and seeing if we can measure the results." So it's kind of a black box thing; we don't know exactly what's going on but let's see if we can do this. Let's see if we can measure on the other end what comes out here and see whether it is efficacious in the same way that we do with many of the other more conventional treatments that we have.

And it's really medical history. That's how medicine has always advanced. If you get real about reading our medical history, a lot of times we didn't know exactly how or why the treatments we were using on people were effective, we just measured and found out they seemed to be helping people.

Doran: *...I want to go immediately to your remark about medical history. And a lot of your listeners may already know this story, and you may know this story, but the thing about energy, energy therapies, is that they are invisible. That's the most difficult thing. That is why moving forward in this highly valuable therapeutic direction is so challenging for us because if it is invisible, people don't think of it as something valuable or useful as opposed to taking an aspirin or something like that. If you are used to something physical from the medical community, then you can't imagine that something invisible could be helping you.*

Now, the example I want to use occurred [in the 1800s]...at this time there was an extremely high mortality rate of women during childbirth... there was a physician [Ignaz Semmelweis] who noticed that doctors did not wash their hands after handling one patient; they just went on to the next patient... and he decided to see if washing hands would make a difference. So he went out of his way absolutely every single time to wash his hands after going from one woman who delivered her baby to another. And the mortality rate in his hospital plummeted once he started doing that.

[But amazingly] he was a laughingstock within his hospital and within his profession. People made fun of him; they thought he was crazy. They thought he was ridiculous. Well, some years later the microscope was invented and people could see the microbes on their hands. And so what all of those doctors for all that time were doing was transferring infected matter from one person to another to another. Women got septicemia and all kinds of things and died.

This was a doctor who just moved forward with a theory that he thought would work. It was proven but he became the laughingstock of a lot of people... But it changed. It changed the course of medical history. Now every doctor washes their hands. They use Latex gloves. They use all kinds of sanitary precautions and so the death rate has dropped dramatically in all kinds of ways. But if it wasn't for that one guy who said, "I can't see anything on my hands." And the doctors would say, "Look, my hands are fine. There is no blood on it or anything." Well no, you can't see what's really on them.

So that is what I say the state of energy therapies is right now. Also, I must tell, these kinds of therapies are far more accepted in other countries than in America...

Later:

Me: *...let's shift gears for a minute here because what we're actually going to do, and I'm very excited about it, I am actually going to get a series of therapeutic treatments from you to help me with my condition. And I am going to pay you for that, as would be* *appropriate, but I am going to share some of that process with folks as I go through it and what the results might be.*

So for me, I really would like to do the Bengston therapy because of some of the reasons we've talked about here and because I have had a chance to talk to Dr. Bengston and hear his story for myself. I do have a health issue that I want to address, and at the same time, I want to explore this more fully because of the implications for the larger questions of consciousness and our relationship as human beings to extended consciousness. So with that, let's shift gears for a minute and why don't you tell me, as someone who is preparing for a Bengston therapy, what you would tell me and what I need to know, and ask any questions that you would ask.

Doran: Great, very good. So when someone says, "I would like to do Bengston therapy," there is [often] a very long [series of] emails that [describe] Frequently Asked Questions about Bengston therapy that we send to them—so we will be sending you that.

There are a couple of things to know in advance. First of all, it seems to us, and I am going to give a slightly broader foundation here, there is a whole branch of medicine... called psychoneuroimmunology... this is the relationship of thoughts and emotions to physical disease. And pretty much everybody, even in the straight, strictly western medical community, now understands that 80% or 90% of physical diseases have an underlying emotional component.

And so the first thing the Bengston method does is collapse whatever the emotional infrastructure is, whatever the emotional scaffolding is, that might have allowed the biochemical platform for disease to manifest. It is very, very common—I would say 99.9% of the time—for people who are receiving Bengston therapy to have two things happen. First, they have tremendous emotional releases immediately, like the day after the treatment. You may feel suddenly overwhelmingly sad, suddenly overwhelmingly angry, all kinds of things for no apparent reason. The reason is the Bengston method is working on that emotional scaffolding first because the physical disease is unlikely to be strongly helped unless that emotional scaffolding is gone. It will just come back.

The other way people experience the release of that emotional infrastructure is through dreams. It appears to us that dreams are an extremely efficient evacuation route for emotions, and a lot of people, especially the first few nights after the first treatment or two, will have dreams about longstanding emotional set points—people who are very fearful have a lot of fearful dreams. People who are angry have a lot of hostile dreams. They are very interesting, and we are complicated little music boxes, way more than we think about on a day-to-day basis. But the Bengston method helps us kind of slice and dice what all these elements are.

So we tell you to expect that. We also tell you to expect to feel the need for much, much more sleep than usual. We are not sure why that is ... We also have several clinical cases where people's belief systems—it's not that they don't believe in the therapy, but they have limiting belief systems that we feel get in the way of the energy helping. And so we believe that the therapy also makes you sleep because then your mind doesn't get in the way and undo what the therapy is trying to do. So we would tell you about that.

Me: *Okay, and just to interject again, I don't have to believe any of that, right? I don't have to accept that is going to happen. I don't have to try and make that happen. I can be skeptical that is going to happen. I can say, "Gee, this probably won't work for me," which I think most people in this situation would say. All of that is okay, right?*

Doran: *Oh, yeah, absolutely without question—100% okay. You do not need to believe anything. However, I will tell you part of the way of generating the therapy, if you are trained in this therapy, is something that Bill [Bengston] developed early on called image cycling. Now, if you are interested in helping yourself as we are trying to help you—and you do not have to do this, you do not have to do this at all—but if you are interested there is something called image* *cycling...*

Note that I never learned or used image cycling or any other "positive thinking" technique. I just imagined I was a mouse in a cage.

Doran: *Let's see—what else? And the other thing, we generally prefer to do the therapy at night simply because you will probably fall asleep within five or ten minutes of my starting the treatment. If you want to take a nap during the day, that's totally fine, we can set it up, but we always ask you to be lying down, whether you are on the couch, whether you are in bed, or whatever. We just don't want you to be operating heavy machinery.*

Me: *We will be able to do this over the phone, right?*

Doran: *Yes, although it's not required. We don't need to be on the phone. We can be on the phone but a distance healing does not require a physical connection like Skype or the phone or anything like that. Absolutely, distant healing means that you just lie down and I send you the energy.*

Me: *That sounds pretty simple. So I tell you what, I am not going to go into any specifics on the condition right now, but I will share that information with you via email and then I will share it with our listeners after this whole therapy is done and I report on the results. So I think that we will just keep it really simple.*

So, while I was waiting for the cardiologist to accept my new insurance, I started doing energy healing sessions with Bernadette Doran. At the same time, I began exploring other alternative treatments for my condition. I found an excellent website called StopHeartPalpitations.com and, after doing my own research, followed many of their dietary recommendations. So as the story continues, there were two kinds of treatments going on at the same time; one was this dietary change, and the other was this energy healing. I realize that was confounding treatments, but I didn't give a heck; I was just trying to get better. I was really concerned about my heart and I was feeling terrible. I was determined to do whatever I could to feel better.

So, as I researched and implemented some dietary changes and vitamin supplements, I began doing energy healing with Doran in parallel. It was rather amazing. The first time we scheduled a healing, I laid down about two minutes before the healing was to begin (9 p.m. my local time). I almost immediately felt a strange, warm, painful sensation right in my heart. Of course, you would think I would associate this with the healing, but I didn't. I looked at the clock and it was exactly 9:01. It was then that the connection struck me. The sensation lasted for a few more minutes and then it just kind of went away. I spent the rest of the hour in a somewhat meditative state before drifting off to sleep.

The next day I felt terrible, worse than before. My condition had escalated: heart palpitations, heart pain, and a general feeling of an upset stomach. I simply didn't feel well. I reported this to Doran, and she immediately informed me that this is a very common condition. She said I should expect to feel worse for a while after the treatments and then feel better. Over the next couple of days, I did feel better.

She also told me to notice my dreams. During the first night, I didn't notice anything particularly different about my dreaming. However, the second night I had some very deep, profound dreams. One could argue this was somehow planted by suggestion, but I had some dramatic, profound dreams that felt were related to my treatment.

I followed through with eight energy-healing sessions in all with Doran. About a week after the eighth treatment, I finally reached a point I hadn't been in months: I didn't think about heart palpitations. They weren't there; they were gone. I have had a couple of recurrences in the months since, but only a few, and very mild ones at that. For me, the treatment was an amazing success.

I don't know if anyone else would have had the same experience. I also don't know how much was due to Bernadette's treatment versus the change of diet. But time and time again, during the period when she was supposed to be sending me energy, I felt a sensation. I felt a definite shift in my health.

I still haven't completely absorbed this energy healing treatment into my worldview. I have absolutely no idea how a person 2,000 miles away can use a very structured, ordered technique to channel energy to someone. I don't know how, but I know it defies a mind-equals-brain understanding of human consciousness.

CHAPTER 10:
WRONG ABOUT ATHEISM...

The atheist assumption is that there is nothing beyond the physical. That our consciousness, whatever we call the soul or consciousness or [whatever], it's nothing but atoms and energy interacting according to the laws of physics...that it's all just physical...The [analogous] assumption they are making [is] that truth is entirely encapsulated within formal [mathematical] systems...but [Gödel's theorem] has demonstrated this is not the case...it hints at a metaphysical component of consciousness...it demonstrates their assumption...[the belief] that [the materialistic worldview] can entirely encapsulate truth, is false.[36] —Physicist Matthew Pulver

Lee Mitchell and I were going to fight. It was the first day of school, and we had decided, without input from anyone else, that the title of toughest seventh-grader had come down to one of us. We would square off on the front lawn of the unfortunate folks who lived at the corner of Spring Avenue and 51st street—a block from our suburban middle school.

The battle was an uneventful draw, with each one of us unable to land any hard shots, but the buildup was quite a scene. I still remember the large crowd that marched down the street to see the brawl that fizzled.

Atheists, too, always seem ready for a good fight. Whether on television or the internet, atheists always seem ready for conflict. But the usual battle between atheists and their religious rivals often obscures the strange, symbiotic relationship between atheism and status-quo science.

Atheism isn't just about religion bashing—although atheists do plenty of that. It's about offering a science-centered alternative to the Christian-infused worldview they see in Western culture. The Christian paradigm insists there is more to this world than meets the eye. The atheist counter-punch is the kind of mind-equals-brain scientific materialism that inevitably leads to them to the conclusion that life is a meaningless illusion created by biologic robots. Even though many atheists don't realize the full implications of this silliness, it has become the default position in their battle against narrow-minded religious conservatives.

While I've taken some hard shots at organized religion on *Skeptiko*, I saved most of my ire for the atheistic nonsense permeating academia. Among my many interviews with academic atheists, three from the fall of 2013 stand out. Two of those interviews, one with Oxford philosophy professor Dr. Stephen Law and another with seminary-student-turned-Christian-debunker John Loftus, were the result of a surprising exchange I had with New York University professor of psychology Dr. Gary Marcus.

Dr. Gary Marcus

Although Marcus appeared on *Skeptiko* to promote his bestselling books and provocative *New Yorker* pieces on consciousness (*Skeptiko* #217), our interview sounded like a lot of the debates I was having with hardcore atheists.

Dr. Gary Marcus: *I'm also very, very skeptical of [near-death experience science]. It doesn't make sense to me to be honest. It doesn't fit with the rest of the stuff that I understand about how the brain works, which leads me to believe that something is likely be-*

ing misinterpreted. I can't promise that and I haven't read every word on it.

Me: *Have you read any word on it? Have you read any of the leading [researchers] out there?*

Dr. Marcus: *I've read a few words here or there but it doesn't make sense to me. It would be like you asking me have I read anything on astrology. I mean, I know about astrology but I don't see the causal mechanisms.*

Me: *Yeah, but it's not really astrology. You've got Parnia at Cornell; you've got the University of Virginia researchers. You've got a lot of well-respected people who've studied it for a long time and are publishing... Dr. Bruce Greyson and all those folks...*

Marcus was being coy. Earlier in the interview he had already revealed his hand. He answered one of my questions about human consciousness by saying, "Whatever the answer is, it's biological." Now he was being evasive about near-death experience research. But he soon demonstrated his bias by likening NDE research done at some of the world's leading hospitals to the study of astrology. Well, I can be coy, too. I titled this episode of *Skeptiko*, "Dr. Gary Marcus Sandbagged by Near-Death Experience Science Questions." As if a NYU psychology professor who regularly writes articles on human consciousness could be "sandbagged" by basic questions about the survival of consciousness after death.

What struck me about Gary Marcus' interview was how much it sounded like the Darwin-thumping, fundamentalist atheists I'd talked to. Not only were the talking points the same, but the thinly veiled zeal was a perfect match as well. I expected Dr. Marcus to give a nod to status-quo, atheistic psychology, but not this level of commitment. After all, it's not like Marcus was asked to wave the flag for atheism, or make the rounds at atheist meet-ups. He was doing the NYU-psychology-professor-thing, but in the process he was parroting the same talking points I'd heard from hardcore atheists.

Dialoging With Atheists

My encounter with Dr. Marcus bugged me. Some respected academics, like Rice University's Dr. Jeffrey Kripal (*Skeptiko* #115 & #176), had alerted me to the dark, soul-crushing negativism that permeates academia, but now I was encountering it first-hand. I couldn't believe Marcus was swallowing these shallow-thinking atheist talking points. And it upset me to think he was spreading this dogma to all those doe-eyed students at NYU.

I decided to make another round through the Amazon bestseller list looking for prominent atheist authors. I was determined to expose this silliness at its source. My search lead me to two books: Dr. Stephen Law's *Believing Bullshit* and John Loftus' *The Outsider Test for Faith*. Both were near the top of Amazon's list of bestsellers in the Atheism category, and both authors agreed to join me on *Skeptiko*.

I didn't get very far with philosophy professor and outspoken atheist Dr. Stephen Law (*Skeptiko* #219) because, even though he has a PhD from Oxford in philosophy, he seemed ill-equipped to deal with the philosophical implications of scientific-materialism and the evidence that contradicts it.

Me: *...you're also Provost for the Centre for Inquiry in the UK, which I'm not sure how it's perceived in the UK but in the U.S. it's known as basically a debunking organization for all manner of paranormal claims.*

That leads me directly into this last topic I want to talk to you about in terms of philosophy of science. As you were [discussing belief], what kept ringing through my ears is what I hear all the time, this idea of extraordinary claims require extraordinary proof. I mean, you want to talk about sweeping mystery, sweeping evidence that you don't like, under the rug. Here is the mantra that the whole Centre for Inquiry crowd pulls out...I see that as just an intellectually feeble kind of pronouncement. Extraordinary claims require extraordinary proof—that is anti-science, isn't it?

Dr. Stephen Law: *Why do you think that?*

Me: *We've built this whole institution of science, the whole process of peer-review, the whole process of self-correction around this idea that we will, [by working together], discover what is real, what is not real; what is extraordinary, what is not extraordinary. So the idea that after the fact, after the results come in, we say, "You know, that's pretty interesting result, but I deem that to be extraordinary; therefore, you need an extra level of proof on that." I think it's just silly.*

Dr. Stephen Law: *Okay, I think I see where you're coming from. The way I've understood that principle, extraordinary claims require extraordinary evidence, says that suppose I tell you... I've got a mobile phone and a cup... You're going to go, "Hey, yeah, that's good enough for me. Steve's got a mobile phone and a cup." If I now wield out a fairy which I make dance on the end of my finger and go, "There you go, a fairy on the end of my finger," you're going to go, "Yeah, Steve's got a fairy on the end of his finger. Fair enough. I'll accept that on the basis of the same kind of evidence that I accepted he's got a cup with a mobile phone." I bet you would not.*

Me: *Sure, sure. We're talking about science here. We're talking about peer-review. Here's the example that I sent you and I have personal experience with because he told it to me and we discussed it on this show. That's British psychologist and parapsychology critic Richard Wiseman, who has investigated probably more of these paranormal parapsychology claims, like telepathy, than just about anybody else. Here's his quote: "I agree that by the standards of any other area of science that remote viewing [and he later added ESP] is proven. But that begs the question: do we need higher standards of evidence when we study the paranormal?"*

So Stephen, this is not the fairy in the cup. This is a guy who has reviewed hundreds of peer-reviewed papers and is saying, "You know what? It's good enough for any other field, but not good enough because of the ground-breaking upset it would make for sci-

ence." This is the best evidence I could give you for my claim about scientific materialism being woven into science-as-we know-it.

The inability of academics like Dr. Law to grasp the paradigm-busting potential of consciousness science is hard to comprehend. It's one thing to have a spirited debate over whether near-death experience science proves consciousness survives death; it is another to debate why it matters.

Part of the reason atheists are such sloppy thinkers when it comes to big picture science is they've wasted much of their intellectual muscle debunking Christianity. They act like falsifying claims of virgin births and empty tombs gives them a free pass when examining whether consciousness is an illusion. This intellectually lazy approach was at play during my interview with author and Christian debunker John Loftus (*Skeptiko* #221):

Me: *What I really focused on is falsifying this "biological robot" claim, this mind-equals-brain idea. It's falsified all over the place, and I've got dozens of very well qualified, highly respected researchers who suggest that's the case.*

One of the main topics we talk about on this show a lot is the published research on near-death experience. A lot of people don't know that there are over 100 peer-reviewed scientific papers on near-death experience. I'm wondering—have you looked at near-death experience science at all? Have you dealt with it on your blog? I know it's not in your book because I looked there. Any thoughts on that?

John Loftus: *Well, Victor Stenger wrote a chapter in my anthology,* The End of Christianity, *where he looked at Dinesh D'Souza's claims of those sorts of things...*

Me: *Hold on. Victor Stenger's been on the show (Skeptiko #191). I don't think he fared very well. I don't think his chapter on near-death experience holds up. He doesn't know what he's talking about. And Dinesh D'Souza is not a near-death experience researcher. He's a Christian Apologist.*

So one of the things I think a lot of atheist folks don't understand is when we talk about NDE science we're talking about people at universities who are publishing in peer-reviewed journals. These are not people who have a preconceived idea or agendas they're driving toward. These researchers are simply saying, "This patient got up off the table after being resuscitated and said—this is what I saw."

Anyone would be curious about investigating this... so they have, and this is what they found out. We have to deal with this evidence one way or another.

Loftus: *Oh, I agree. All I can do is point you to the experts. Again, that's not my focus. Keith Augustine and Michael Martin, the atheist philosopher. Keith Augustine is the Executive Director and Scholarly Paper Editor of InternetInfidels.org. They've done a lot of research on those sorts of things and I think...*

Me: *No, they haven't, but that's okay. It's not your thing, I get it. But those people haven't done anything.*

Loftus: *You know what? I have a predisposition against those sorts of things because I have no clue what the difference between what a mind and brain is. I don't know where the mind is located. I don't know how the mind works.*

Later in the interview Loftus claimed that my questions about consciousness had "blindsided" him. This led to an interesting email exchange. First, I invited him back for a "do over." He declined. Lotus then sent me a rambling email in which he called "my God" "irrelevant and unnecessary." A strange claim given that the term "God" was never mentioned during our interview. Lotus then proceeded to tell me how this survival of consciousness was not a "big deal." I felt obligated to offer a different opinion:

Me: *I think you've kind of missed the thrust of our discussion so let me re-cap the three mains points:*

1. Atheists, even popular media savvy ones like you who write books and give public presentations, often make claims (e.g. "we are all biological robots," "life is meaningless," "consciousness is an illusion") without realizing the implications. You demonstrated this in our interview when you seemed genuinely confused when I pointed out that your love for family can be nothing more than a meaningless illusion (according to your belief system).

2. Most atheists live their lives in a way that is inconsistent with their stated beliefs. I think you do care about your family. I think you do act as if your choices matter. I think you do think you are doing good in the world by countering the negative impact of fundamentalist religions. But during our interview you seemed unable to resolve the absurdity of your, "bite the bullet, we're all just biological robots" with the life you live.

3. "New Atheists" are in a stuck-on-stupid debate with fundamentalist Christians when the real action is whether or not consciousness is an illusion. If consciousness is more than an epiphenomenon of the brain (as the data suggests), and if consciousness survives bodily death (as the data suggests), then we have to take seriously questions about the order and meaning of consciousness. You don't have to be an expert in consciousness science to see that this leaves little room for your brand of New Atheism...

If WE survive death, then WE are not an illusion... WE are not biological robots. This changes everything for the New Atheist crowd, as well as for materialistic science in general.

I don't seek out these contentious dialogues, but I don't shy away from them either. I respect atheists like John Loftus for "seeking truth," I just wish they'd look a little harder.

One more thing, several of my guests have charged me with "sandbagging" or "blindsiding" them during interviews. This invariably happens after they've failed miserably during the interview. This smokescreen tactic is usually transparent, but I don't always have the opportunity to prove how silly these claims are. I did in the case of John Loftus and subsequently published the original email I sent when I invited him on *Skeptiko*:

Wrong About Atheism...

Hi John... I'd like to know if you'd be available for an interview to discuss your blog and recent book, The Outsider Test for Faith.

While I've enjoyed many of your withering attacks on goofy Christian Apologetics, I'm always left wondering whether the science of human consciousness has been brought to the table.

Do people have genuine spiritual experiences? Is our mind purely a function of our brain? Are these topics/questions you've examined and would you be open to discussing them on Skeptiko?

Blindsided, indeed.

Atheists don't have a Bible, but if they did it would be Charles Darwin's explanation of the theory of evolution in his book, *The Origin of Species*. Next, we'll examine how the history of this foundational pillar of science-as-we-know-it is radically different from what we've been told.

CHAPTER 11: WRONG ABOUT EVOLUTION... AND CHARLES DARWIN

Darwinian Theory had a profound impact on perception of man's "animal" nature. In the late nineteenth century, "social Darwinism" was used to justify cutthroat business practices, and even today a "survival of the fittest" attitude prevails in many areas of society. At this point, one can only wonder about the possible effect of a [alternative] theory which emphasizes the cooperative aspects of evolution and depicts us as intimate participants in the fate of all of nature.[37] — Physicist F. David Peat

One of the things I like about parenting (my wife Joni and I have four children) is helping with homework. I wasn't a great student, especially in my early years, so I've enjoyed helping my kids while revisiting what I skimmed over the first time around. It's also a great chance to see culture making in action. After all, what we put in our textbooks reflects who we think we are.

The other day my youngest son, Zane, asked for help with his upcoming science exam on evolution. I was excited. I've done several shows on Charles Darwin and Alfred Russel Wallace so I thought I might be able to help. I was about to give Zane a full initiation into the controversy when I caught myself and realized that nothing I had learned was going to be on his exam. So I stuck to the textbook for studying, and I told him I was writing a book chapter that might explain the real story.

The theory of evolution is a lightning-rod issue for science. The idiotic battle over whether Adam and Eve or chimps are responsible for the current state of affairs is beyond silly. But while the theory of evolution has clear and direct implications for consciousness science, *this chapter is not about the theory of evolution*. This chapter is about history. It's about the stories we tell ourselves and how they've come to define our culture.

One way my experience producing *Skeptiko* has changed me has to do with how I approach what are commonly called "conspiracy theories." I'm now much more open to them. Looking back, I'm not sure why I wasn't more receptive to begin with. My experience as an entrepreneur had taught me to look for "conspiracies" in every deal because I had found that every important business transaction involving a considerable amount of money resulted in a conspiracy of some sort. For example, if you want to sell your high tech company to one of two competitors—you're creating a conspiracy. If you want to buy a 100 acres of land in an area rumored to be the site of the new airport—you're walking into a conspiracy. Conspiracies run the world of business, but I naively thought that was "just business." *Skeptiko* showed me a different reality.

If you're turned off by conspiracy theories and like to stick to the "everybody says so" version of history, then you're probably going to find this chapter challenging. But, if you're open to looking at the best evidence for how the theory of evolution came into being, and how its origin impacts our science-as-we-know-it driven world, then please read on.

Michael Flannery

Evolution is real. Our world is a reflection of it. Or, maybe it would be more accurate to say, we *choose* to live in a world created by Darwinian evolution. Perhaps the choice has to do with worldview. Here are some excerpts from three *Skeptiko* interviews that changed my understanding of this history.

Most of us have come to believe we are competitors in a battle for survival. My interview with professor Michael Flannery showed me there's more to this story. Flannery is Associate Director for Historical Collections at the Lister Hill Library at the University of Alabama, Birmingham, and author of *Alfred Russel Wallace: A Rediscovered Life*. He's an expert on Charles Darwin and the lesser known, but widely acknowledged co-discoverer of the theory of evolution, Alfred Russel Wallace. Flannery has written a book many consider one of the best biographies on Wallace. He's also reviewed just about every important book on Darwin and other key figures from this time period.

Through my interview with Flannery, I learned about an important distinction between Alfred Russel Wallace and Charles Darwin in developing the theory of evolution. Wallace understood that evolution isn't about survival of the fittest as much as it is non-survival of the least fit. You don't have to win the race; just don't come in last.

This may be self-evident once someone points it out, but it embodies a huge paradigm shift for our "winning-isn't-the-everything-it's-the-only-thing" culture. Wallace saw this. Darwin did not. Wallace also saw (probably because he, rather than Darwin, discovered evolution...but more on that later) that evolution isn't about survival of the individual, but survival of the group. Any advantage an individual gains is meaningless unless it improves the survivability of the group. Life, it would seem, is really about cooperation over competition.

Here are parts of my interview with Professor Flannery (*Skeptiko* #149):

Me: ... *I really have enjoyed learning about some of the won-*

derful things you've discovered about Alfred Russel Wallace. The breadth of your knowledge is really impressive... But I have to start with the real question that we all want to know, Mike. How do you think Noah got all those dinosaurs on the boat?

Professor Michael Flannery: [Laughs]

Me: I'm kidding, obviously. But, I feel like I have to start there. And it's frustrating because in researching this show I've listened to really smart scientists like astrophysicist Neil deGrasse Tyson and [Richard] Dawkins and I've heard them beat this crazy Creationist-scare-tactic thing to death. And so as I ask that question tongue-in-cheek, but I also want to slip in a real question—do you think the situation would be different... would I still be asking that ridiculous question if Alfred Russel Wallace's ideas about evolution had won out?

Professor Flannery: That's a very good question and I think the obvious short answer is, no. You know, Wallace had his own view. I guess it might be helpful for your audience to sort of step back and review what Alfred Russel Wallace really did because certainly by comparison to Charles Darwin, he is more-or-less the unknown entity.

So let me just start off by saying that Alfred Russel Wallace was the co-discoverer of natural selection, so he was very much involved with the early thinking along the lines of evolutionary theory. In fact, he's the one who actually prompted Charles Darwin to rush his work to press after he had received Wallace's letter from the Spice Islands.

However, Alfred Russel Wallace soon diverged or departed from Darwin's theory in proposing what I have referred to as "intelligent evolution," which is actually directed, detectably designed and purposeful common descent. In other words, there is teleology involved in certain aspects of evolution.

Later:

Professor Flannery: When you look at the dividing lines and you start looking at Charles Darwin's theory, Charles Darwin was—and I think this is the key ingredient to understanding Darwin's theory—Charles Darwin's theory was really a mechanism, I think, that really was designed to support and explicate naturalism. More specifically, methodological naturalism and the idea that a scientist must invoke only natural processes functioning via unbroken natural laws in non-purposeful ways.

Me: So are you saying he started with naturalism and then looked for the mechanism to explain our world from a naturalistic viewpoint?

Professor Flannery: I believe he did. I believe he did because he had already been introduced to materialism as a philosophy as a 17-year-old member of the Plinian Society at the University of Edinburgh. He knew William Browne, William Gregg; he became a close companion and confidante of Robert Edmond Grant, who was, I believe, 16 years his senior but was a thorough-going evolutionist and a thorough-going materialist.

Me: Hold on, because this is really an interesting point, and it's one of these subtle points that a lot of people are going to miss because the traditional story regarding how Darwin came to discover the theory of evolution is more along the lines of him being this good Christian guy who's marching along with everyone else and is then overwhelmed by the evidence he receives on the Galapagos Islands. He has this "aha moment" and he says, "Oh my gosh, as hard as it is for me to accept this, as a scientist, I know this must be the way we all came to be."

Professor Flannery: And Alex, I can understand why people believe that because that's essentially what Darwin wanted us to believe in his autobiography. That's exactly the way Darwin sketches out how he came to his own theory of evolution. However, when you start digging and start looking into his life, you find out that that doesn't quite match up with his own autobiographical statements.

In fact, when we talk about Robert Edmond Grant, [Adrian]

Desmond and [James] Moore did what I think is a very good biography on Charles Darwin. It came out in the '90s... They in effect more-or-less were supporters of Darwin, so they don't have an axe to grind, but they make this interesting comment. I think it's very important to understand this.

They refer to Charles Darwin and Robert Edmond Grant's meeting as "decisive." They say, "Darwin was coming under the wing of an uncompromising evolutionist. Nothing was sacred for Grant. He saw no spiritual power behind nature's throne. The origin and evolution of life were due simply to physical and chemical forces all obeying natural laws." Now this is the person that the young, teenaged Charles Darwin would go on quiet, personal strolls with through the countryside.

Later:

Me: *And what about Wallace?*

Professor Flannery: *Well, it's different. It's very different. In fact, you have to understand something about Wallace and Darwin when you look at them as historic figures. Charles Darwin was pretty much a made-man. I mean, he was wealthy. He came from a family of doctors. His father, Robert Darwin, had been a very successful physician.*

His grandfather before him, Erasmus Darwin, who by the way, wrote one of the early transmutationist texts called, Zoonomia, *which came out in 1794, which Charles Darwin read avidly and took notes on, was also a physician. So Darwin really never had to work. That's neither here nor there, but it actually does come into play as to how they developed their evolutionary theories. At any rate, Darwin did not have to struggle for an existence.*

He didn't have to make a living, per se. He had good investments; he was a member of the upper classes of Victorian society. So most of what he did, admittedly he engaged in it with a passion, but it was, by comparison to Wallace, a hobby.

Wallace came from a different side of the tracks. He came from a family that was in what we might call middle-class decline. He had no formal education other than attending the Mechanics Institute of London for a while, in stark contrast to Darwin's Cambridge University. He spent some time as a land surveyor with his brother, but then gets a botany field manual and is absolutely captivated by nature. So he becomes this self-taught botanist and then zoologist.

He decides to go off with Walter Henry Bates, a friend of his, off to the Amazon. He spends four years in the Amazon, from 1848 to 1852. He comes back; he has a terrible experience there. His ship sinks, it catches on fire and he's at sea for about 10 days before he's picked up. He loses all of his specimens and everything, so he never really completes his work. He winds up looking for another place to go. He then spends eight years in the Malay Archipelago, from 1854 to 1862.

So all this to say the field experience of these two naturalists is also very different. Darwin has five years on the H.M.S. Beagle, 1831 to 1836. [With] Wallace you've got 12 years of field experience, four years in South America in the Amazon River Basin and the Rio Negro and Uapes River Valley, and eight years in the Malay Archipelago, where he goes across scores and scores of islands in that island chain.

Me: *Why do you think that's important, Mike? What can you put your finger on where you see where that crops up in terms of...*

Professor Flannery: *Well, there are a couple of things. It made the collecting very different. When Darwin collected, you have to ask, what's Darwin doing? He's basically skirting the coast with Captain Fitzroy, okay? He has time, at leisure, to examine individual species at great length, and that's not a bad thing. That's a good thing. But he is under no specific pressure to amass or accumulate large numbers of specimens.*

Wallace did. The reason Wallace did was he was sending those back through an agent in London and selling them. This was how he was making his living. So if he didn't amass enough specimens to return, he wasn't going to put any food on the table. It also forced him to collect twice. He had to collect once for himself and once for his agent in London, okay? So by the time he gets done, he amasses over 125,000 specimens. Darwin doesn't amass anything like that [number]...

Later:

Me: *...So can you come back and in layman's terms explain the differences between Darwin's view of competition and Wallace's view of competition?*

Professor Flannery: *Wallace tended to view competition occurring among groups in a demographic sense. Darwin tended to view it as individual competition.*

Flannery on Dr. Jerry Coyne

After learning about the Wallace-Darwin controversy, I couldn't understand why this history wasn't better known among evolution enthusiasts. I had the chance to interview one such enthusiast, University of Chicago professor Dr. Jerry Coyne (*Skeptiko* #161). Coyne has cultivated a reputation as a hard-nosed, in-your-face atheist who's determined to stamp out the horrible Creationists he sees lurking in every corner. But, in my interview with him, I was dumbfounded by how he seemed ignorant of the well-established history of Wallace and Darwin. This excerpt from my follow-up interview with Professor Michael Flannery (*Skeptiko* #162, which begins with a recap of the previous episode) will show you what I mean:

Me: *Now you'll recall that at the end of the last episode of* Skeptiko *I told you I was going to do this interview because when I was talking to Dr. Jerry Coyne he had a number of things to say about this relationship between Darwin and Wallace, and in particular about whether or not Alfred Russel Wallace ever connected biogeography to evolution. This sounds like "inside baseball" and detail-oriented stuff you may not care about, but it turns out to be pretty central to this culture war debate surrounding the theory of evolution...*

So I've managed to get Professor Mike Flannery on the phone here and Professor Flannery was nice enough to actually review the interview that we had with Jerry Coyne when I sent it to him. I thought there were some direct points about the Darwin versus Wallace thing that he certainly knows a lot better than I do. I thought we'd have Professor Flannery back on here. Mike, thanks for joining me.

Professor Flannery: Sure.

Me: *So just a couple of quick points here. I feel first of all that we have to frame this up. Dr. Jerry Coyne says, "Where was I wrong? You tell me where I was wrong." So I feel we have to oblige Dr. Coyne in that regard. Where I want to start, because I think he describes you as not a Wallace scholar but as an "Intelligent Design Creationist" and I think when someone does that it slips into another mode, it's not at all scientific... he's calling you a Creationist to discredit you. It's funny to me because when we started our last interview on* Skeptiko, *we joked about "how many dinosaurs fit in Noah's Ark?" We were laughing about how these discussions quickly degrade into silliness, and yet here he is with the Creationist thing. What did you think when you first heard that?*

Professor Flannery: Well, it's sort of when you don't have a real point to make, sometimes the only point you can make is the ad hominem argument. You attack the person rather than the idea. The very first thing I thought is, "Well, I've written a book about Alfred Russel Wallace. Whatever else may be said for it, the book should stand or fall on its own merit."

Me: *Let's go over it a little bit. I think you, as you demonstrated in your book and your interview with me, have an extensive knowledge of not just Wallace but of many, many of the Darwin scholars, Darwin biographers. You impress me as someone who's quite well read and published on this topic. So, to call you a Wallace scholar I think is pretty accurate.*

Professor Flannery: Well, I thank you. I think the book has a real contribution to make and I've spent a considerable amount of time studying over the past four or five years Alfred Russel Wallace. If you look at the book, Alfred Russel Wallace: A Rediscovered Life, it has well over 200 references in it and it's thoroughly documented. Again, I would just say I believe the book will stand on its own merits.

Me: *I think it will, too. Let's get to some of the specifics where we can point out some questionable areas in Dr. Coyne's presentation. These are direct quotes. Here's quote #1: "Wallace did not use biogeography as evidence of evolution. I mean, never."*

Professor Flannery: *He seems to really be unfamiliar with Wallace's body of writing on that topic. As I point out, Wallace himself actually pointed out, that this is not just a matter of applying some arbitrary divisions of the earth that might be useful to the zoologists but in fact it's a fundamental problem with the Theory of Evolution. His two-volume work,* The Geographical Distribution of Animals, *which he published in 1876, was written specifically to address that.*

Now if that's not using biogeography as evidence for evolution, I don't know what it is. Let's be quite clear that Wallace's reputation as a biogeographer extends beyond the seminal work in that field, and that's the one I just mentioned. His work dates back to 1852.

Me: *Give us that date again because another quote from Dr. Coyne is, "Wallace never made an argument until much after Darwin," and later he says "...17 years after 1859."*

Professor Flannery: *Again, he's referring to* The Geographical Distribution of Animals, *which is the magnum opus in the field. There's no doubt about it, but actually the very first paper was actually a work that Wallace read before the Zoological Society of London. He actually read it in December of 1852. It was a paper called* "On the Monkeys of the Amazon," *and he was then just beginning to be looking at geographical distributions of various species. This is in 1852, so this is well before 1859 when the first edition of* Origin of Species *is published, okay?*

Then in 1855, [Wallace] writes during the rainy season and he was on the island of Sarawak and he writes a paper called "On the Law Which Has Regulated the Introduction of New Species." This actually appeared in The Annals and Magazine of Natural History. *It appeared in September of 1855. [It dealt with] the geographical distribution of animals in the Malay Archipelago, which is where he had been spending a considerable amount of his time. In fact, he spent a total of eight years there. He was applying the geographical distribution of animals in that region to evolution.*

Flannery went on to give many more examples of how Coyne, who is a leading flag-bearer for Neo-Darwinism, lacks a basic understating of the history of Wallace and Darwin. At this point, you're probably not surprised to hear about another well-credentialed academic stumbling in their defense of science-as-we-know-it. But, Coyne's failure with biogeography was nothing more than a warm-up to what I was about to discover—the best evidence suggests that Darwin plagiarized Wallace in order to develop his own theories.

Roy Davies

Roy Davies is a former BBC filmmaker and journalist. My interview with him (*Skeptiko* #247) gave me a fascinating new perspective on the Darwin-Wallace interplay I had explored in previous interviews.

Me: Roy, this book of yours is one of the most underappreciated books I have run across. The implications of your theory, if they are true, and I think they are, are dramatic. So let me let me lay out a thumbnail sketch of your theory. You claim that Charles Darwin, that God among Gods of science...

Roy Davies: (Laughs) Yes, he is a demi-God.

Me: While he was toiling away on this theory of evolution, he received this gift, this wonderful boost, in the form of a letter from a guy who was also working on the theory of evolution, a guy who is acknowledged as the co-discoverer of the theory of evolution, Alfred

Russel Wallace. And you claim that Darwin plagiarized, stole, some of the key elements of his theory from Alfred Russel Wallace, and then that he lied about it. Is that the gist of your claim?

Davies: Yes.

Me: Okay, simple enough. Tell us a little about your background and how you [came to be] interested in this part of history.

Davies: Okay, right, let's start at the beginning. My primary training was as a journalist, a journalist on a newspaper in South Wales in Britain. And [one] of the most important lessons that I learned there was from a senior reporter who said to me, quite simply, "Look, from now on, do not accept any story that you are told by less than two people." Meaning if one person tells you a story, you go out and check it with somebody else who doesn't know that person and make sure it is the right story. And that stayed with me as a pretty good line about how to approach stories.

Eventually I left journalism and went back to college. I joined the BBC and after some years I became the managing director or the managing editor of history programs at the BBC. And one day... one of my producers to came in to see me. And he said he wanted to make a program about Charles Darwin. A new book had been written, this was 1990, 1991... and he wanted to make a program about it. I didn't know much about Charles Darwin, except the theory of evolution, and I wanted to learn as much as he knew. So I said, "Okay, it's a good story, let's make it." And he came back later with a very good program which went out on the BBC under my name as editor... and that program was called The Devil's Chaplain...

A few years later, I had left the BBC, and a woman producer from Wales came up to me and said, "You made the Darwin program." And I said, "Yes." And she said, "You told the wrong story." I was totally flummoxed, and I said, "What do you mean I told the wrong story? That's the story everybody knows." And she said. "Yes, of course it is. It's the story everybody has been told and everybody has bought into, but it's not the correct story. In fact, you never mentioned Alfred Russel Wallace." And that hit me because I thought, "Well, who the hell was that guy?"

And so from that moment on, I started researching Alfred Russel Wallace and what I found flabbergasted me. I discovered one thing after another came out regarding the papers and documents of the time; none of which I expected. It led me to write [The Darwin Conspiracy].

Me: That's quite a story; I love that. The journalist whose journalistic sensibilities are challenged by someone who comes up and says, "You didn't get the story right." And then you have the integrity to say, "Gee, if I didn't get the story right, I have to fix it." So before we talk about what you discovered, the evidence for it, and what some of your critics have said—which I think is very important and I love to do that on this show...

Davies: By all means.

Me: Let's talk about why this is important because a lot of people would love to make this about religion, baby Jesus, the Bible, Creationism. That's where a lot of critics, especially, Roy, folks here in the States, they would like to take it there because it's an easy argument to win. They can just label you as some kind of Creationist, some guy who has some religious agenda. So let's get that out of the way.

Davies: My stance personally is that I have no religion. I am totally an irreligious person. I have no belief at all in the afterlife, God, creator, whatever, none. This is simply about two men. This is about two men who, more or less at the same time, in different parts of the world were trying to do the same thing. One, here in Britain, the other floating around the world in different locations. And, it's about the connection between these two men. The connection between these two men who never met during the years they were trying to prove the same thing. They only met afterwards. And it's also about the holes between them, including the 8,000 miles that separated them. So I became so interested in this story and I just couldn't let go of it. The research and writing took me 12 years to complete and resulted in the book you have in front of you.

Me: Great. Let me add that this story, and the way this competition between Wallace and Darwin plays out, is important because Darwinian evolution has become the lens through which science sees the world.

[On Skeptiko], we have spent a lot of time investigating issues around human consciousness because if you want to look at big picture science questions... you always come back to the question of consciousness: "Am I a biological robot? Is there a 'me' in there? Is my brain just creating this illusion of me?" These questions are at the bedrock of every question science asks. And Darwinian evolution is really the foundational pillar in science's answer to those questions, the position that concludes, "You are just a biological robot and a meaningless part of nature's machine..." So whatever we make of that, whether you buy into that or not, we have to realize this is why it's an important part of history. It's important because that's what Darwinian evolution has become.

Davies: Absolutely.

Me: ...We have a reason to look at this a little deeper, and ask some bigger questions about why this matters, about why we should probe so deeply into the kind of questions that you raise.

I would like to have people continue [to ask], "Why does this matter? Who cares?" And then [ask] "Why do I live in the world that I live in?" And I think hopefully at the end of this road we can come back and re-examine some of those questions after we look at, most importantly, the evidence. We have to find out if you really did discover anything, if this theory of yours makes any sense, and how it stands up to the critics. So let's start with the letter. What is the letter?

Davies: There were two letters and, once upon a time, I thought that I might call this book, "Two Letters for Mr. Darwin" because the whole point of this story revolves around these two letters. It revolves around when Alfred Russel Wallace made his discoveries and when Charles Darwin claimed those discoveries as his own.

[So], there were two different letters—one was sent in October 1856 and that included Wallace's own ideas about the basic tools, if you like, of evolution. And the second one was sent in March 1858, and that became the famous letter that Darwin claimed he received two weeks later than he did. He made this claim so he could fill up 66 pages of his ongoing thesis, the natural selection manuscript he was writing.

Me: His journal, right?

Davies: No, it wasn't his journal. This was the natural selection manuscript that he was then completing. And in that manuscript he introduced 66 pages of information about the theory of divergence, which Wallace had already sent him. Darwin then copied it into his natural selection manuscript and claimed he had it there for some time. Then he announced he had received a letter from Wallace, which in fact he had in his own collection for about two weeks. It was the letter that became the famous letter which started the whole Linnaean procedure in London on the first of July, 1858. It was there that Wallace and Darwin were given the accolade of being the discoverers of the theory of evolution. But Charles Darwin's name was put first. Hence, although Charles Darwin and Wallace are both the discoverers, eventually it became Charles Darwin as the discoverer of evolution and Wallace's name was forgotten.

Me: Okay, I want to pour over the details of some of that because it's obviously very important. First, I want to talk about this two-week difference. You alluded to the fact that we can trace a significant amount of material Darwin adds to his body of work during that two-week period. How do we know that?

Davies: Okay, what he wrote in those 66 pages only came out when Origin of Species was published in 1859. Before that... nothing was referred to in terms of these 66 pages of new information. Darwin didn't present those pages to [Joseph Dalton] Hooker or

[Charles] Lyell in his case for claiming the theory of evolution as his own. Lyell and Hooker never knew that had taken place; they just were not informed.

And so it was important because Darwin knew he had these 66 new pages of Wallace's information taken from, I believe, Wallace's own outline of his theory... That's the plagiarism charge.

Wallace sent the letter to Darwin completely innocently. He just wanted Darwin's approval over the fact he'd come up with the theory of evolution. And when Darwin got the letter I think he panicked, started writing it into his manuscript, and a few days later claimed to Hooker that he had a competitor to his theory. What he didn't tell Hooker was that it was Wallace's idea. He said that it was his own.

So divergence became the thing which allowed both Lyell and Hooker to believe that Darwin came up with the theory of evolution on his own. And they also believed Darwin when he told them he was going to lose out because Wallace had a similar idea. That's why the two-week delay matters. It was, and the evidence I have offered proves it, Darwin at the lowest ebb he had reached.

But, a long time before, he had started doing this sort of stuff to other people at different times. It wasn't the first time that he cheated.

Me: *And even your critics, while they might not agree on that last point, that Darwin had a history of fudging with his sources, there is agreement on the fact that this is either an incredible coincidence or something more. It's a coincidence that Darwin has this explosion of ideas—announces to his colleagues that he has had this incredible breakthrough—during these two weeks that are in question about when he received the letter from Wallace.*

It's also an incredible coincidence, if we're to believe the status quo traditional story, that Wallace would be this co-discoverer right alongside Darwin who was kind of puttering around with these ideas, that they would both focus in on this idea [and that it] would crystallize for both of them at exactly the same time. Given the length of time that it takes for this correspondence to go back and forth, it's either an incredible coincidence or it's something more: It's plagiarism, as you have suggested. Can we at least boil it down to that? I mean, we do have to agree that there is no other way to explain it.

Davies: *I would say that the story of [coincidence] is just too much to believe. And Darwin's own history in this area is suspect. For people to claim that this was a coincidence is just outlandish. There is no point, and I make this point in the book, as do other academics—there is no point at which Darwin's research come before Wallace's ideas. Wallace's ideas, from the time that he goes to South America and then eventually on to the Malay Archipelago, he is sending ideas back which Darwin then finds a way of getting into his own research. It never happens the other way around.*

Me: *Right, although some of your critics have suggested that it did happen. One of your critics said that Darwin even writes in his journal about Wallace's paper and he says "nothing new." The point being there is nothing new here from Wallace.*

Davies: *Yes, that's absolutely true, and they were writing about the Sarawak [paper] in September 1855, which Wallace had published in* The Annals and Magazine of Natural History. *And basically what Darwin [claimed] in the marginal notes [was] that there was "nothing new here." And to show that there was "nothing new," he actually put a straight marginal line down against 35 of Wallace's paragraphs in his own copy of that [paper], indicating thereby that he found each of those paragraphs of interest... Why would he indicate the paragraphs had some significance? Each one of them, 35 different ones in that piece.*

Me: *And I think, as you alluded to just a minute ago, I think the strongest circumstantial evidence is the rush to publication, right? Here is a guy who has been dallying around with this stuff for years, hasn't put it together, and now he gets this letter and all of the sudden there is this rush to get it to publication. And the fact that he brings Wallace along as the co-discoverer sounds more like a guilt* exercise than anything else.

But as you point out, Wallace seems to be somewhat naïve about what's going on. In his letter to Darwin he seems to be saying, "Hey, you're an upper crust kind of guy with all the right connections, maybe you can help me get this published." And he is handing it over to a guy who says, "Oh my gosh, here is this stuff." And then there is this tremendous rush to get it to publication, is that right?

Davies: Well, the story has been told before about Mozart and Salieri, and it's almost exactly the same story. Somebody who wants something so desperately and yet somebody else has it; how on earth do you go about getting that for yourself? And Darwin managed to do it. He managed to convince people, because he was upper crust, and because he and Wallace had exchanged letters before. And Wallace didn't know anybody in the upper crust section of British society, and certainly not in scientific society, except his own agent, Samuel Stephens. And it was Stephens who kept on sending him notes saying, "Will you stop theorizing please. Everybody here wants you just to collect things and send them back." And of course, Wallace's whole instinct was [on these bigger questions]: "How on earth does life change in each generation? How do the generations change inside a particular species?"

And it was [Wallace] who actually found all this and put it together. And then because... he sent Darwin the letter, Darwin replied saying to him, "Well, we thought much alike." In fact... they had not "thought much alike," and yet Darwin had convinced Wallace that they "thought much alike." Wallace had his own suspicions... but then he sends another letter back to Darwin. Then Darwin replies and the last letter of the five of them was the one where Wallace actually says to him, "Look, I have come up with the theory of evolution. Would you have a look at it? And if you think it's good enough, would you show it to Charles Lyell?" And that's where the story kind of gets bogged down in the whole question of what was in that letter because [the letter] doesn't exist anymore.

Wrong About Evolution... And Charles Darwin

Charles Darwin made sure the letters don't exist anymore, not only to Wallace but to all the significant characters who might have offered some kind of criticism of what he was doing, particularly people like Asa Gray. None of Gray's letters to Darwin... exist at this moment, but Darwin's letters to Gray [do exist] and in those letters [Darwin] gives the game away.

So, we know for a fact that there is something seriously wrong with this story. I mean, people who love Darwin and are Darwinists keep on saying, "But you can't prove it, you can't prove it." And that's partly because Darwin destroyed the evidence which might have been able to bring him down. But that isn't the only evidence we have. We have got a lot of evidence, mostly that Wallace's ideas were first [criticized] by Darwin and then taken and made into his own.

Me: Right, and we should also mention, as long as you're talking about it, what letters meant during this time period. Because now we send off an email and we throw it in our trash and don't think anything of it. Letters, letters of this kind, to be lost—to be thrown away—it's virtually unheard of for a scientist like Darwin, right?

Davies: Indeed. And Darwin keeps all the letters, like from his son going to the university... He keeps those letters, but the significant ones are not there. The letters themselves, you see, and this is the most important part—everything in this story hinges [on the letters] and this is the difficult bit [to get] across to people. Everything in the story hinges essentially on two things. One is the mail service between Singapore and [London]. When it happened, how it happened, which ships took the mail, which ships were meant to take the mail, and which ships actually carried the mail [back to London]. The second part of it is the whole idea of letters themselves and their significance in the middle of the 19th century. The whole idea of letters coming from an area of the world where the British were in trouble in China with the Chinese uprising... [and in trouble with] the whole Indian mutiny which had begun in 1857, which had been boiling since 1856. And those two things made the mail from Singapore to home the most important channel of information for anybody governing the British, in India or in China.

Me: Okay, so let's put an exclamation point on that. Mail service is the internet, the railroad, the highway system, all wrapped together for the British Empire. It's the way the whole thing runs back then.

Davies: Absolutely.

Me: And most importantly, what you reveal in this book and through your research is that there is no way that Darwin could have gotten the letter on the day he said he did. It really doesn't make any sense. And as a matter of fact, even your critics agree it would be a highly unusual set of circumstances for it to arrive on that day. And the reason is because, as you just said, the mail at this point in history is a regularly scheduled event. And when it comes from the other side of the world, we can trace down when it happened. So, why do you think it is so unlikely that Charles Darwin received that letter on the date that he said he did?

Davies: Okay, there is no likelihood at all. One of my critics is John van Whye, who is now in Singapore, [and is] a Darwin enthusiast. Now, the point that John van Whye is making is that, John, because he needs [the delivery of the letter] to happen in the middle of the month, [he developed a theory that] gets the first boat to pick up the mail and take it to Singapore, totally without precedent.

Me: ... If there is a better example of apologetics, I don't know where it is. If there is a better example of Dr. John van Whye starting with the conclusion that he wants and then working backwards to try and make it fit the circumstances... tell us why he is jumping through all these hoops... the regularly-scheduled mail, the mail that gets reported in the newspaper around the world, would not put the letter there at the time that Darwin claims that it is there, right?

Roy Davies: Let me come the other way around. The mail coming back from the Far East, from Singapore, [arrived in] two deliveries a month from Singapore to Britain. There had [been one per month

Wrong About Evolution… And Charles Darwin

in 1856], but the British, because of the connection with India and China, were in desperate need of having a more regular service to find out exactly what was happening. They didn't know what their Army was doing; they didn't know what was happening in the Indian Revolution and the mutiny. Nobody had any idea until the mail came in. So… the biggest headlines in the Times of London every two weeks was—the mail has arrived. It was the most amazing piece of news and it always led the papers on that day. Now then, the mail that came from Malay Archipelago, where Wallace was at the time he discovered the theory, always came from the Archipelago to Britain and arrived on the second day or the third day of every month. The second delivery [of mail from the Far East] always ended up in the middle of the month, but that middle of the month mail never included mail from the Archipelago. Now, John van Whye insists that the mail from the Archipelago arrived on June the 18th and I am saying that because of what I know of the schedule of boats in the Far East at that time, that letter posted on the 9th of March 1958 arrived on June the 2nd in London, on June the 3rd in Darwin's home. And it was for that reason that for the next two weeks Darwin had the chance to introduce into his natural selection manuscript the 66 pages of new information that he had taken from Wallace's own theory.

That is why he was able to say to Lyell in his letters, "Look, this is appalling. I have worked on this for years and here I get letters from Wallace saying exactly the same things. His ideas could stand as headlines to my chapters." Or words to that effect.

The theory of evolution shouldn't matter as much as it does. Evolution is true enough; it seems to be a reasonable approximation of how living organisms change. But do we really need to elevate this tiny bit of history to the untouchable status it has among many scientists and committed atheists? Does it really answer our deepest questions about who we are and where we came from? Or is the theory of evolution protected so fiercely because it's a vehicle for propping up our absurd science-as-we-know-it, mind-equals-brain paradigm? Given what Davies and Flannery have discovered, I think we should demand a script re-write.

CHAPTER 12:
THE END OF SCIENCE-AS-WE-KNOW-IT

Science fails when it insists too strongly upon Fact, Truth, Knowledge, or aligns itself with a social order that is fundamentally hostile to change and simply treats science as a pimp treats a whore.[38]
— Author Curtis White

When I graduated college in 1981, I had an undergraduate degree in Information Systems and an MBA. At the time, accounting firms like Price Waterhouse and Arthur Andersen were scooping up fresh-faced grads, dressing them in Brooks Brother's suits, and billing them out to their Fortune 500 clients as IT Consultants. I interviewed with several such firms and eventually took a job with what is now Pricewaterhouse Coopers.

But I really wanted a job with Arthur Andersen. I remember my lunch interview with a team of Arthur Andersen consultants at a hip downtown Chicago eatery. The interviewers were probably only a couple of years older than I was, but they still managed to make me feel like a high school freshman at the senior prom. I flubbed the interview and didn't get the job.

At the time of my interview, Arthur Andersen was the number one accounting firm in the world with thousands of employees worldwide. I don't think anyone could have imagined that in 2002, they would surrender their license to practice accounting after a handful of auditors were found guilty of criminal charges relating to the Enron scandal.

Sometimes change is sudden; the collapse of Arthur Andersen happened in months. Could we be at the brink of such a change in science? Is the materialistic, science-as-we-know-it paradigm at a tipping-point? Are thousands of go-along-to-get-along scientists ready to loosen up their lab coats and move past this absurd insistence that we are all biological robots in a meaningless world? As you've seen, there is plenty of evidence in favor of a more expansive view of who we are, but do we have the will to embark on such a massive shift?

I asked philosopher Bernardo Kastrup, PhD (*Skeptiko* #248) and UCLA research psychiatrist Dr. Jeffrey Schwartz (*Skeptiko* #250) these very questions:

Dr. Bernardo Kastrup

Me: What are we talking about when we talk about scientific materialism? What is it?

Dr. Bernardo Kastrup: *Personally, I'd like to split these words. When people say, "scientific materialism" it's a philosophical, an ontological interpretation of science. Science itself is just a study of the patterns and the regularities we observe. It doesn't carry with it an interpretation in terms of saying, "Oh, this is outside mind [but this in not]." It doesn't matter for pure science.*

Scientific materialism is when you load the scientific observations of nature with an ontological interpretation, and you say that what you're observing here is outside of mind, that it is something that has an existence that would still go on even if nobody were looking at it. That is already an interpretation. It's not really pure science anymore. The essence of scientific materialism is that the real world is outside of mind.

Me: What you say in [Why Materialism is Baloney]—and I like the way you put this—you say it's our fundamental metaphysical assumption. I think that's going to catch a lot of people by surprise because those are not words we generally associate with science. Yet, as you point out, our current paradigm has done this sleight of hand, where they're taking metaphysics and substituting it for science. So take what you just said and break it down for us in terms of why you think that metaphysical assumption is absurd.

Dr. Kastrup: Everything that we can ever know is an experience. It's generated in our mind, it's our, it's a subjective experience. So we can break that down with a living example. The computer I have in front of me, the chair I'm sitting on, the table I feel with my hands, the lights in the office where I am, these are all experiences of color, shape, smell, and sound. These are all experiences, they're all in mind and that's the only reality we can know.

What scientific materialism states is that all this experience is created by your brain and is therefore all inside our head. The real world, they say, has no color, no sound, no shape. In that sense, materialism is an incredibly strong metaphysics—strong in the sense that it says reality is not what you experience every day. They claim it's just an epiphenomenon, something you create inside your head.

Me: I have interviewed some very well respected academics who hold very firmly to this materialistic view of consciousness. There's a certain fallback position they use when they realize there's an absurdity to it. I'd like you to talk about this emergent property of the brain idea, where we're still 100% brain based but somehow we introduce this idea of an emergent property of the brain.

Dr. Kastrup: Yeah, I think it's an appeal to magic. It's a label for something that we don't know and can't even conceive, so we give it a name to make it sound like we understand it. Emergence is a phenomenon that exists; for instance, sand dunes are an emergent phenomenon of grains of sand and wind. They form beautiful patterns, and when you look at those patterns, they are unexpected. How can just grains of sand and wind form those patterns that look designed? But these are all examples of what philosophers call "weak emergence."

Basically, weak emergence is when you have a phenomenon that's surprising compared to its components, like in this case of sand and wind, but which is still explainable by its components. You can still get the properties of what you're seeing from the properties of its basic components, like grains of sand and wind. You can simulate sand dunes on a computer and get the same beautiful patterns.

What people mean when they call consciousness an emergent phenomenon of the brain is an appeal to what philosopher Dave Chalmers calls "strong emergence." That's when the phenomena you observe cannot be deduced from the properties of its components. So, in the case of consciousness, it can't be deduced from the properties of its components. Consciousness cannot be deduced from spin, momentum, or the mass of subatomic particles. You simply cannot deduce the redness of red, the pain that we feel when we lose a loved one. You can't deduce that phenomenology from spin, charge, momentum, mass, and whatever we have in the material world. So that's an example of what philosophers would call strong emergence.

I'm of the opinion that strong emergence isn't coherent. It's an appeal to the unknown. It is an appeal to magic. It's not only an unknown; it's something that we can't even conceive. There's a gaping black hole in our view of reality. We've given it a name and we say we will find out about this in the future. It looks all cozy and warm, and we get that fuzzy feeling as if we are honest there, but in fact there is a universe to be bridged here—one that will never be bridged because it's incoherent. It's based on the fundamentally flawed interpretation of reality. That's the position I take.

Dr. Jeffrey Schwartz

Kastrup is a philosopher who looks at materialistic science from the "big picture" perspective. Dr. Jeffrey Schwartz is a research psychiatrist at UCLA School of Medicine who works with patients suffering from Obsessive Compulsive Disorder. Both have come to the same conclusion about the limitations of science-as-we-know-it.

Dr. Jeffrey Schwartz: *...When one has a deep ideological commitment to materialism, which we all know is the reigning worldview of our current era, certainly among academic elites, and certainly among academic elites in neuroscience, [then one is] committed ideologically to the notion that everything is the brain. So you get into this profound mismatch between [the] mind-brain understanding you need to apply [in] day-to-day living and this profound ideological commitment to the belief that everything is coming from the brain. It's a philosophical/ideological mismatch between the kind of worldview you need to actually live in the day-to-day world, and this huge philosophical/ideological structure that the elites of our era feel committed to defend.*

And [then there are] the funding agencies that provide the funds for research, etc., I mean, there is a social structure that insists that this materialist [worldview] be adhered to. It's a prerequisite for playing the game, for having a laboratory, for doing high-level research in our era. So this mismatch exists. And a lot of people are willing to say that it's not really working out in the clinical realm. I mean, the clinical application of that materialist paradigm is not doing well.

Me: *What about people in academia who see this mismatch, who have to say, "[materialism] is a wonderful suit" because they can't say the emperor has no clothes. How do we get past that?*

Dr. Schwartz: *It's funny because there's a huge amount of historical precedent for situations like this. The human drama really is the playing out of this kind of situation... where worldviews become entrenched. Obviously, this is what's happened in our era.*

Science is extremely good at explaining material aspects of reality. So it came to pass that they came to believe they should be good at explaining all aspects of reality. Science in the way it's done to explain material aspects of reality does not explain human behavior particularly well. It's radically incomplete.

Does it have a contribution to make? Definitely, it has a contribution to make. But I think it's fair to say that any reasonable person would know that the science that works so well in explaining the material world does not work nearly so well... explaining how human beings act, and what human beings are, and what it means to be a person, a human—a living, breathing, and even to use the word spiritual, human being in the real world.

So, you know, "twas ever thus." That's really the point. I mean, this is always the case. There is always a mismatch between the worldview that is supported by the economic powers-that-be in any era and how people live their lives.

Let's be clear. Only one side of the mismatch has a professional, economic commitment to supporting materialism. I mean, it's the people who make their living supporting that perspective. So it's not so hard to explain why they are deeply committed to that perspective because they have a profound economic interest in that perspective. Otherwise, you're left with these very deep questions about why one perspective becomes the elite perspective in any given era.

As I just said, I think in this particular case you can understand it because of the success science had in one realm, the material realm, that lead to a general belief that science is going to give us all the answers. And since science works in the material realm, then we have to sort of make-believe that everything is the material realm, because that's the realm that science works in.

Me: *There's a lot of talk about paradigm change and "consciousness being primary" and us coming to that understanding. Do you really envision that ever happening, given what you just described as the huge economic incentive, or at least economic* infrastructure, that's been built around the existing paradigm we have? Let's get real, can we ever envision moving away from this?

Dr. Schwartz: The answer is definitely yes, over a long period of time. I mean, I'm 62, do I expect to live long enough to see this change? Candidly, I don't. But I have a lot of friends in their 20s, so it's interesting.

Our life spans are getting longer. So, for people in their 20s now who 100 years from now will be 120 and might well be alive at that age, I think there's a reasonable chance that they're going to see something like that happen. I mean there's probably going to be a lot of strides in the next 100 years, but I guess I can conclude by saying; look the advances in the research of ... "mindfulness." I've been working with it for 44 years and there has been genuine advance in research on the subject of mindfulness. Some of it is very, very good; a lot of it is mediocre. But the advances being made are definitely consistent with people coming to deal with the fact that the choices and decisions that one makes about how to focus attention have real effects on the brain. Even people in academia are doing that research. You can see the top researchers, people who have been engaged in this kind of mindfulness research for over 10 years now, who used to be staunch materialists, they're having second thoughts. And they're saying, "Huh, there seems to be more to this than just the brain because the brain alone doesn't explain the data." Things are improving at the margins. There are definitely reasons to be positive.

Dr. Schwartz along with many of the smartest thinkers I've spoken to offer hope for our future. But do we really have to wait?

CHAPTER 13:
SCIENCE OVER THE TIPPING POINT

Science is predicated on the concept that science is arriving at truth—a unique truth...In a way science has become the religion of the modern age. It plays the role that religion used to play of giving us truth...However, they find that they can't do that...not only because scientists share what everyone else shares—assumptions and opinions—but also because the very notion which has been defining science today is that we are going to get truth. But that may not be a valid assumption because thought is abstraction, which inherently implies limitation...So we can see that there is no [ultimate] "road" to truth.[39] — Physicist David Bohm

I was a slow reader. I've gotten better as an adult, but as I child there was nothing I feared more than a lengthy reading assignment. However, like most kids, I loved having books read to me, especially fairy tales. One of my favorites was *The Emperor's New Clothes* by Hans Christian Andersen. As you probably remember, it's a story about two swindlers who promise a vain Emperor the finest suit of clothes from a fabric invisible to anyone unfit to see it. The Emperor's advisors see the con being played out but pretend to see the Emperor's clothes out of fear of losing their positions.

Finally, the swindlers do a fake fitting and the Emperor marches before his subjects. The townsfolk are afraid to speak up for fear of the Emperor's power until one boy in the crowd blurts out the obvious: the Emperor has no clothes.

This is the position science-as-we-know-it finds itself in. The absurdity of "you are a biological robot living a meaningless existence" is self-evident. It fails philosophically and has been falsified experimentally. Most importantly, we, the townsfolk, know it's not true. We don't live our lives as if we're biological robots because everything about life tells us we're not. But in the halls of academia, and in the respected journals of science, the myth must be propped-up. To stay in the respectable-science club, one must nod as the Emperor marches by wearing nothing at all.

Let's end this charade. Let's tell these delusional scholars we don't want to play this game any longer. Let's, with a loud, clear voice, call for an end to this silliness.

We have to demand more from those who seek to be intellectual leaders and yet fail to show the courage to stand up for the truth-seeking ideals of their profession. The tools of science are a wonderful and powerful gift. They offer the potential to allow us to rise above our biases and prejudices and see new vistas. Properly applied, science might even give us clues about what we should do with the precious little time our life provides. But none of this is possible if we accept the dehumanizing absurdity of science-as-we-know-it. We must set ourselves free from the socially engineered political joke science has become and open ourselves to what lies beyond this narrow thinking.

Tim Freke

My interview with philosopher and consciousness explorer Tim Freke from June 2014 (*Skeptiko* #249) captures the challenges and opportunities of moving beyond science-as-we-know-it. During the interview, he had used the word "mystery" several times to capture the complexity of big picture science questions, so I asked him: "What do you mean by mystery?"

Tim Freke: *...We're having this experience. There are these colors, and shapes, and thoughts, and feelings, and we don't know what it is. As you enter that "don't know," consciousness changes. Then the mystery is not just the not knowing; the mystery is knowing [there is] mystery...*

Me: *I also like how you contrast it with the non-mystery that we're so used to, and that is forced upon [us]. So, you talk about reductionistic science, it's all about non-mystery. I like how you work in this idea of "It's just..." That is the ultimate counter to mystery, "Oh no, no, it's just..."*

Tim Freke: *Yeah, the "just" word. It really struck me, exactly what you've just said. I love science. I love all human knowledge, and I love science. I love its attitude. I love what it shows. I hate reductionism, because of this narrowness where it reduces the richness of life to one narrow story. You can always say, "It's just..." [Someone says:] "Oh, the birds are singing." [And we hear:] "No, that's just them holding their territory." Well, it is them, holding their territory, but it's not "just" that. Or, "I've fallen in love." [We hear:] "Really, that's just chemicals..." Well, it is chemicals, but it's not "just" chemicals. And it's seeing [that this is always the case] no matter how much we understand of life. When you look at the sunset, [you] still go "WOW!" And, that "wow" is the mystery. When we lose that, whether it's looking at the world in terms of our understanding of it, or just in our relationship with our lover, when you lose [that], the mystery goes dead on you. And my whole relationship with life feels a bit like that. It's a lover. And, when I lose the mystery, it goes dead on me. I put it in a box. I think I know what it is, and I don't. I need to remember that for it to stay alive.*

Me: *Tim, what about this deep need we have... for non-mystery, for certainty, for understanding, because that's what they're playing on. So, in the same way that the scientists are playing on that [and]* *saying, "Oh no, it's just..." the atheists are playing on it and saying, "It's not..." And the fundamentalist religious people are playing on it and saying, "Oh no, it's just this..." Even, as you quite fairly point out, [there are] folks in the non-duality community playing the game too, and saying, "It's not. It can't be more. It can't be God in the sense that you've thought about it. It must not be that." So what is it about our need for non-mystery, our need for certainty, that makes us so susceptible to this line of thinking?*

Tim Freke: *For me, Alex, the key... is what I call a paralogical approach. Paralogical thinking, which is a word I've coined to capture a tradition which is very, very old actually, is to understand that the nature of this mystery of life that confronts us is inherently paradoxical. [This] means, as the great physicist Neil Bohr said in his idea of complementarity, "To understand the nature of life, we need to two opposite but complimentary perspectives at once." If we don't see it with both eyes, we don't see any depth, if you like. So, we need both of these. So, yes, I'm a spokesman for the mystery, but I love words, and I love story. My whole job as a philosopher is to create the best story I can. And I need that. I can't live without a story. If I lived just in the mystery, I would be an amnesiac. I couldn't function; it would be ridiculous. Even these noises I'm making would have no significance because I would have no memory of language, and so on. I need story, but I don't want to get lost in story. I need not this or that but both...*

Later:

Me: *... I've heard your presentations, you do a wonderfully expansive, inclusive, and yet at the same time discerning deconstruction of our culture and how we deal with these things. And by that I mean you're inclusive of science; you're inclusive of religion, but you also deconstruct them. I wonder sometimes if you're a little bit too inclusive of science. Too complimentary of, "Oh, science is so great." And then you make this distinction of, "Well, there's* **deep** *science and then there's this reductionistic, materialist science..."*

Science Over the Tipping Point

Whenever we qualify and say, "Well, when I say science, I mean this... when I say religion, I mean this," I think we're kind of confusing the issue a little bit. Science is science-as-we-know-it. Science is science as it works in our culture. And that science is disturbingly myopically reductionistic. It's soul crushing in so many ways. So do we really need to embrace science, or do we need to take a step back and... clean out the wood shed a little bit? [Don't we need to] say there are some real problems, not only theoretically and philosophically, but experimentally, with this idea of reductionism?

Fundamentally science is about [this idea] that we can measure things; that we can be certain about things... We've known that's not true. Ever since quantum physics, and chaos theory, and consciousness science, they all [have told us], you can't really measure things at all.

Tim Freke: *That was a fantastic little rant, Alex, I love it. Yes, I agree with you completely. You're probably right. I'm far too polite...*

What's been obvious to me over the last period, since finishing my last book... is exactly what you're saying: that a new hegemony has arisen in the West. It's in America; it's in the UK over here. It's a very dead form of science...

That science... would become associated with something it's not... just as religion becomes associated with something it's not... [that] we would end up with this fundamentalist, narrow literally, as you say soul destroying idea... which creates this meaningless idea that we're clever monkeys clinging to a rock hurdling through space on a meaningless journey to nowhere—[even though] I don't see anything in the findings in the experimental nature of science which points to this at all. I think what's happened is the way of investigating the world has been conflated with a philosophical interpretation of [it]. And the philosophy of reductionism and materialism is dead. It is finished. It doesn't work. And yet somehow that very stupid idea has taken off as if it's the way that we should see reality now, and it's disastrous. I think it is completely disastrous. And if I've been too polite in the past, I'd like to take the opportunity now to not be polite and say— I really think the job today is to undercut it, because it's in danger of suffocating us.

Consciousness Is Not Meaningless

Science is wrong about almost everything because science is married to this absurd idea that you and I don't really exist. It says we're just an illusion created by this meaningless electrochemical reaction going on inside our brain. The denial of our consciousness as anything more than the product of a biological robot turns out to be a big deal not just philosophically but experimentally. If we can't account for the effects consciousness might have in our experiments—if we can't measure consciousness—then we can't really measure anything.

Science-as-we-know-it gets around this problem by declaring consciousness meaningless, but it's not. You are not meaningless. Your conscious experience isn't meaningless. And the feelings you have for the most important people in your life are not meaningless. This doesn't mean I know what the meaning is—I do not. But I'm on a mission to discover what it means for me.

The *Skeptiko* interviews I've compiled have changed me. They've turned my world upside down more than once. But the knowledge I've gained has made me a better husband, father, and friend. I've discovered and re-discovered myself again and again and, in the process, I've gained a deeper connection with those I love and care about. Knowledge is power, and sharing knowledge, like so many of my guests on *Skeptiko* have done, is the ultimate gift one person can offer another. I've avoided conclusions and advice in this book because I don't have much to offer, except this: *Follow the data, wherever it leads.* That's what I'm going to do.

APPENDIX:
Dr. Dean Radin's Meta-Analysis of Presentiment—by John Maguire

When parapsychologist Dr. Dean Radin appeared on *Skeptiko* in 2007 (*Skeptiko* #2), he was asked to respond to the following quote from University of Oregon psychology professor Dr. Ray Hyman:

Dr. Ray Hyman: *What's fascinating about Radin is that he comes through as being very sophisticated... I've gone back and looked at a lot of other stuff he's done in the ESP field, and everything he does he has some new, novel application of technology, the latest in computer sophistication, and so on. And yet he gets results and then he never repeats that; that's dropped and he goes on to other stuff. And as a cynic you wonder, well why? What happened here? Why isn't he able to follow up on his own "great work," right?*

Then I found out some other things. When I went through his presentiment work, I found—it took me a while to find it—that he's done three experiments, each replicated the other. But when I go down to the fancy way that he was correcting for baselines and stuff like that, the correction was one way in one experiment. The second experiment was a different correction. The third experiment was a different correction.

Then I realized...I did some simulations, and the corrections from the first experiment and the ones from the second experiment would cancel each other out. *In other words, it would give a different result. And why was he always changing his corrections?* [Laughter] *And I'm realizing that this is a subtle, kind of, maybe unconscious, not conscious so much, way of making sure you're going to get what you want from the data.*

Hyman starts out with some unfounded personal attacks, then hits hard with charges of academic dishonesty by claiming Radin was "*changing his corrections.*" To be specific, Hyman claims Radin is fiddling with the "baseline." This deserves an explanation. The research in question has to do with an unusual phenomenon known as "presentiment." Dr. Radin suspected human beings might possess the innate ability to sense an event is going to happen *before* it actually happens. He then carried out a series of experiments to test the validity of this hypothesis.

Radin asked test subjects to stare at a blank screen and wait for an image to be displayed. All the while Radin measured how far, and at what times, their physiological responses deviated from the "baseline" measurements. The baseline condition can be thought of as a participant's "normal" physiological state (i.e. physical, mental, emotional, etc.) when no experimental variable or stimulus has yet been applied to the test subject. As the experimental variable or stimuli is applied, the participant's responses continue to be carefully monitored.

This procedure is well tested and commonplace in the field of psychology. It allows experimenters to observe behavior before, during, and after applying stimuli, and lets them gauge how far from the baseline condition test subjects deviate during an experiment. Radin would measure the patient's response over thousands of trials, compare it to their baseline, and assess whether there was a significant, above-chance effect going on.

Hyman accuses Radin of adjusting the baseline measurements of his test subjects to distort, or amplify, the results. He accuses Radin of data-mining or "cherry-picking" data in an effort to make

Appendix

results appear more statistically significant.

But Radin has a powerful counter to this claim—his meta-analysis. After completing each of his three carefully controlled, lengthy (each took over a year to complete) presentiment experiments, Radin analyzed his data. Each time he sought to improve the statistical analysis of the results. As mentioned previously, Radin is a highly competent researcher with a proven track record in academia as well as with some of the world's leading commercial laboratories. He knows his stuff.

After the third experiment, Radin used his latest and most refined methods of data analysis to re-analyze the data from all the experiments taken together. This combining of data from different experiments is called meta-analysis. And while such work can sometimes be complicated, in this case, because the experiments were so similar, and because Radin had run each one, the positive results of the meta-analysis were very revealing. They showed that the effect was robust and is present in all three of Radin's experiments.[40]

Radin's "mini" meta-analysis across his presentiment experiments obliterates Hyman's charge. Applying one method of determining the baseline to all trials, across all experiments, rules out any possibility of tweaking parameters from experiment to experiment in an effort to produce misleading results.

If anyone is data-mining or cherry-picking, it is Hyman. But perhaps we should not be surprised, as Dr. Hyman is one of the founding members of the Committee for the Scientific Investigation of Claims of the Paranormal (aka CSICOP, or CSI), an organization known for over-the-top theatrics in defense of status quo science.

Alex Tsakiris' interview with Radin exposed Hyman's nonsense, but Hyman was never willing to acknowledge his mistake and/or apologize to Radin for slandering him. Dr. Steven Novella, who interviewed Hyman, and published the interview, was urged to make a public correction on his show, but he never did.

Notes

1 Feyerabend, Paul. (1977). *Against Method.* Brooklyn, NY: Verso.

2 Tsakiris, Alex (Producer). (2007, January 8). *Skeptiko* Episode 2 [Audio Podcast]. Retrieved from http://www.*Skeptiko*.com

3 Radin, Dean. (2006). *Entangled Minds.* New York, NY: Paraview Pocket Books.

4 Nagel, Thomas. (2012) *Mind & Cosmos.* New York, NY: Oxford University Press.

5 Dr. Marilyn Schlitz appeared on *Skeptiko* episode No. 4 and her article "Distant Intentionality and Healing: Assessing the Evidence" can be retrieved from http://media.noetic.org/uploads/files/DistantIntentionality.pdf

6 Dosey, Larry. (1999) *Reinventing Medicine.* New York, NY: Harper Collins

7 Van Lommel, Pim. (2001) "NDE in survivors of cardiac arrest: a prospective study in the Netherlands," *The Lancet, 358, 2039-2035.*

8 Rosenblum, Bruce (2011). *Quantum Enigma.* New York, NY: Oxford University Press.

9 Beauregard, Mario (2012). *Mind Wars.* New York, NY: Harper Collins.

10 Sherrard, Philip (1987). *The Rape of Man and Nature.* Golgonooza

Press.

11   Tart, Charles. (2009). *The End of Materialism*. Oakland, CA: New Harbinger Publications.

12   Van Flandern, Tom (1994). *Dark Matter, Missing Planets, and New Comets*. Berkeley, CA: North Atlantic Books

13   Past interviews with the aforementioned experts can all be found at www.*Skeptiko*.com.

14   Mobbs, Dean & Watt, Caroline. (2011). There Is Nothing Paranormal about Near-Death Experiences: How Neuroscience Can Explain Seeing Bright Lights, Meeting the Dead, or Being Convinced You Are One of Them. *Trends in Cognitive Sciences* 15 (10): 447-449.

15   Holden, J. M., Greyson, B., and James, D. (2009). *The Handbook of Near-Death Experiences: Thirty Years of Investigation*. Santa Barbara, CA: Praeger Publishers.

16   Dr. Bruce Greyson, psychiatrist and professor at the University of Virginia, has been one of the preeminent researchers in the field since its inception, publishing scholarly articles and books since 1979.

17   Recently, this discredited article found its way in a book by noted philosopher of mind Dr. Patricia Churchland as "support" that near-death experiences can be explained away as neurological misfiring. I was not surprised, given that Dr. Churchland is known for her extremely reductive views on the nature of consciousness (a position called *eliminative materialism*). But her reactions to my challenging her on this point, when she finally agreed to appear on *Skeptiko* in 2014, provided a memorable confrontation.

18   The full interview with G. M. Woerlee, along with all the other interviews referenced in this chapter, can be found on the *Skeptiko* website (www.*Skeptiko*.com).

19   Long, Jeffrey, with Perry, Paul. (2010). *Evidence of the Afterlife: The Science of Near-Death Experiences*. New York: HarperOne.

20   Van Flandern, Tom (1994). *Dark Matter, Missing Planets, and New*

Notes

Comets. Berkeley, CA: North Atlantic Books

21 See for example: Schwartz, Gary E., with Simon, William L. (2002). *The Afterlife Experiments: Breakthrough Scientific Evidence of Life After Death*. New York: Pocket Books.

22 On the *Skeptiko* website, along with the full transcript of my interview with Julie Beischel and a detailed account of her research into medium communication, you can also read, or listen to, my interviews with Ben Radford, as well as additional interviews I did with other Skeptics and researchers in the field.

23 Julie Beischel's upgrading of the previous triple-blind protocol she had used with Gary Schwartz to a quintuple-blind methodology is discussed in more detail in the following article: Beischel, Julie. (2007). "Contemporary Methods Used in Laboratory-Based Mediumship Research," *The Journal of Parapsychology* 71: 37-68.

24 For more information on that study, see: Beischel, Julie & Schwartz, Gary E. (2007). "Anomalous Information Reception by Research Mediums Demonstrated Using a Novel Triple-Blind Protocol," *EXPLORE: The Journal of Science & Healing* 3 (1): 23-27.

25 Dick, Philip K. (1981). *VALIS*. New York, NY: Random House.

26 *Skeptiko* Episode No. 97: Rupert Sheldrake & Richard Wiseman Clash Over Parapsychology Experiments

27 Sheldrake, Rupert. (1999). *Dogs That Know When Their Owners Are Coming Home*. New York, NY: Random House.

28 Volk, Steve. (2011). *Fringeology*. New York, NY: Harper Collins

29 Sheldrake, Rupert. (1999). *Dogs That Know When Their Owners Are Coming Home*. New York, NY: Random House.

30 For those of you interested in viewing the experiment online see: Tsakiris, Alex. "Psychic Dog Experiment." *YouTube*. YouTube 10 April 2008. https://www.youtube.com/watch?v=SvOzdqnTE2I.

31 Dr. Clive Wynne, personal communication.

32   For further details see: Wiseman, R., Smith, M., Milton, J. (1998). "Can animals detect when their owners are returning home? An experimental test of the 'psychic pet' phenomenon," *British Journal of Psychology*, 89: 453-462.

33   Wiseman, Richard. (2010). *Paranormality: Why We See What Isn't There*. Spin Solutions Ltd.

34   Jammer, Max (1969). *Concepts of Space: The History of Theories of Space in Physics*. Cambridge, MA: Harvard University Press

35   YouTube Lecture: Terence McKenna: Science Was Founded by an Angel. https://www.youtube.com/watch?v=8X6_0jJbcy0

36   Maguire, John (Producer). (2013, May). *Q-Niverse* Episode 1 [Audio Podcast]. Retrieved from www.jmag0904.wordpress.com.

37   John Briggs & F. David Peat. (1986) *The Looking Glass Universe*. New York, NY: Simon & Schuster

38   White, Curtis. (2013) *The Science Delusion*. Brooklyn, NY: Melville House

39   Bohm, David. (1992) *Thought As a System*. New York, NY: Routledge

40   Mossbridge, Julia & Tressoldi, Petruzzi & Utts, Jessica (2012). "Predictive physiological anticipation preceding seemingly unpredictable stimuli: a meta-analysis," *Frontiers In Psychology* 3: 1-18

Index

Afterlife, 39,42,45-50.60,64,145
Alexander, Eben, 34
Atheism, 28-29,123-131

Baseline (Experimental), 4,169-171
Beauregard, Mario, 20-23
Beischel, Julie, 54-67, 70
Bengston, William, 106-118
Blackmore, Susan, 35-37
Bohm, David, 163
Brain Death, 36

Cardiac Arrest, 34, 43-45, 49
Churchland, Patricia, 12-21
Consciousness, 5-20, 26-34, 43-45, 48, 50, 53, 67, 81-83, 114, 117, 121-125, 128-134, 146, 157-160, 164-168
Conspiracy, 134
Coyne, Jerry, 140-143

Darwin, Charles, 125, 131-153
Davies, Roy, 143-154
Dawkins, Richard, 136
Dennett, Daniel, 21, 28
Dillahunty, Matt, 27-31
Dogs That Know, 79-87

Dopey Science Creed, 26-31
Doran, Bernadette, 114-20

Emergence, 13, 157-158
Energy Healing, 106-114, 119-121
Evolution, 10, 20-22, 112, 131-153

Fenwick, Peter, 12, 34
Feyerabend, Paul, 1
Flannery, Michael, 134-143, 154
Freke, Tim, 164-167

Gordon, Suzanne, 34
Graziano, Michael, 8-12
Greyson, Bruce, 12, 125

Holden, Jan, 34-37
Hughes, Bill, 95, 102-103
Hughes, Marilynn, 68-78, 93, 97, 99, 101
Hyman, Ray, 2-4, 169-171

Kastrup, Bernardo, 13, 156-159

Lilias Yoga, 5
Loftus, John, 124-130
Long, Jeffrey, 34-48

Marcus, Gary, 124-126
Mediums, 53-80
Meta-Analysis, 169-171
Moody, Raymond, 33
Moore, Jim, 93-102

Nagel, Thomas, 5
Near-Death Experience, 11-16, 33-50, 124-128
Nelson, Kevin, 35
Novella, Steven, 2, 34, 68-69, 104

Observer Effect, 19-20
Open Source Science, 80-81, 87

Parapsychology, 1, 3, 80-82, 86, 127
Parnia, Sam, 12, 34, 125
Pigliucci, Massimo, 35
Presentiment, 2, 169-171
Proxy Sitter, 55, 70
Psychic, 11-12, 53-103, 107, 113
Pulver, Matthew, 123

Quantum Physics, 19-23, 64, 167
Quintuple Blind Methodology, 56

Radford, Ben, 56, 60-64, 90-103
Radin, Dean, 1-4, 169-171
Religion, 3, 49, 124, 130, 145, 163-167

Sartori, Penny, 34
Schwartz, Gary, 54-55, 156-161
Schwartz, Jeffrey, 156-159
Science, 1-171
Sheldrake, Rupert, 3, 79-87
Shermer, Michael, 35, 80, 90
Skeptics, 2, 23, 29, 37, 49, 55-56, 60-61, 68-69, 80-84, 90-95, 102-107, 113, 118, 124
Skeptiko Podcast, 3-4, 7-8, 12-13, 17, 20, 26-38, 51-56, 67-85, 90-93, 99-106, 114, 124-146, 156, 164, 168-169, 173-175

Tart, Charles, 26
Telepathy, 79-87, 127
The Emperor's New Clothes, 4, 159, 163-164
The Origin of Species, 131, 142, 147

Van Flandern, Tom, 33, 53
van Lommel, Pim, 11, 16-17, 34
Volk, Steve, 80

Wallace, Alfred Russel, 133-147
Wallace, B. Allan, 7
Watt, Caroline, 35-37
Weber, Nancy, 90, 93, 101-103
White, Curtis, 155
Wiseman, Richard, 82-87, 104, 127
Woerlee, G.M., 35-49
Wynne, Clive, 81

Lightning Source UK Ltd.
Milton Keynes UK
UKOW06f1451110315

247687UK00012B/171/P